Microsystems

For further volumes:
http://www.springer.com/series/6289

Tomi Laurila · Vesa Vuorinen
Toni T. Mattila · Markus Turunen
Mervi Paulasto-Kröckel
Jorma K. Kivilahti

Interfacial Compatibility in Microelectronics

Moving Away from the Trial and Error Approach

Tomi Laurila
School of Electrical Engineering
Aalto University
Otakaari 7B
02150 Espoo
Finland

Vesa Vuorinen
School of Electrical Engineering
Aalto University
Otakaari 7B
02150 Espoo
Finland

Toni T. Mattila
School of Electrical Engineering
Aalto University
Otakaari 7B
02150 Espoo
Finland

Markus Turunen
School of Electrical Engineering
Aalto University
Otakaari 7B
02150 Espoo
Finland

Mervi Paulasto-Kröckel
School of Electrical Engineering
Aalto University
Otakaari 7B
02150 Espoo
Finland

Jorma K. Kivilahti
School of Electrical Engineering
Aalto University
Otakaari 7B
02150 Espoo
Finland

ISSN 1389-2134
ISBN 978-1-4471-2469-6   e-ISBN 978-1-4471-2470-2
DOI 10.1007/978-1-4471-2470-2
Springer London Dordrecht Heidelberg New York

British Library Cataloguing in Publication Data
A catalogue record for this book is available from the British Library

Library of Congress Control Number: 2011944651

© Springer-Verlag London 2012
Apart from any fair dealing for the purposes of research or private study, or criticism or review, as permitted under the Copyright, Designs and Patents Act 1988, this publication may only be reproduced, stored or transmitted, in any form or by any means, with the prior permission in writing of the publishers, or in the case of reprographic reproduction in accordance with the terms of licenses issued by the Copyright Licensing Agency. Enquiries concerning reproduction outside those terms should be sent to the publishers.
The use of registered names, trademarks, etc., in this publication does not imply, even in the absence of a specific statement, that such names are exempt from the relevant laws and regulations and therefore free for general use.
The publisher makes no representation, express or implied, with regard to the accuracy of the information contained in this book and cannot accept any legal responsibility or liability for any errors or omissions that may be made.

Printed on acid-free paper

Springer is part of Springer Science+Business Media (www.springer.com)

Preface

We are a group of dedicated material scientists each with 15–25 years of research and development experience in the assembly and interconnect technologies of electronics. Our experience in chemical and mechanical interfacial compatibility between dissimilar materials dates back to the end of the 1980s. Since the early 1990s with the rise of the electronics industry in Finland, we focused our work increasingly into the challenges of materials, assembly technologies and reliability of portable electronics. During the last decade, our research has expanded towards high performance electronics and microsystems in different applications including automotive and biomedical devices.

We have cooperated with numerous electronics companies, including electronics OEMs and their suppliers, semiconductor companies and subcontractors. In this work, experience has shown that the development methods in this branch are often immature from the perspective of materials science and engineering. Even though much progress has taken place over the years, much of the hardware development across the value chain is still mainly experimental. The so-called "trial and error" method is widely used in this industry.

We would like to emphasize that there is an alternative development route applicable for electronics materials and process development. In this book, we present the methods one needs to master in order to have a better understanding, and control, over materials compatibility issues. Rather than trying to provide all the right answers, which we obviously also do not have, we would like to contribute educating people who are able to develop new technologies and solve problems based on a deeper understanding of materials and their interfaces. Chapter 6 of this book demonstrates how this methodology can be employed in the development of reliable IC, package and board level interconnect technologies.

This book does not seek to be a collection of correct recipes. Such a database combined with the correct interpretations on materials interactions would surely be of great benefit to the microelectronics industry. Such an effort, however, lies beyond the capacity of a single research group—even after experiencing for a relatively long co-operation with the industry—as the variation of potential material systems in microelectronics is massive. Not only the possible combinations of bill of materials, design and loading conditions are endless, but also the speed of development in these technologies is constantly bringing new material combinations for analysis. Therefore, a recipe list would be out of date soon. Further, since microelectronics hardware is composed of layers of dissimilar materials with various thicknesses and interfaces, microstructural evolution a system is highly time- and process-dependent, contributing once again to the system complexity.

This book is intended as lecture material for graduate-level students. But the book is also targeted for engineers in the electronics and microsystems technology industry. In the development work taking place inside these industries, the teams of electrical and mechanical engineers, physicists and materials scientists need to work closely together. We are convinced that by assimilating and applying the principles and methods as described in this book, these multidisciplinary teams will solve new challenges they face in the development of new material systems of future technologies in a fraction of the time they would otherwise need if applying purely empirical, "trial and error" like methods.

Contents

1 Introduction: Away from Trial and Error Methods. 1

2 Materials and Interfaces in Microsystems. 7
   2.1 Levels of Interconnections, Typical Stress Factors and
        Related Failure Mechanisms. 7
        2.1.1 IC Level 8
        2.1.2 Package Level 10
        2.1.3 Board Level. 20
   References 22

3 Introduction to Mechanics of Materials 25
   3.1 Deformation of Electronic Materials 27
   3.2 Restoration of Plastically Deformed Metals 30
   3.3 Formation of Strains and Stresses in the
        Electronic Assemblies 31
        3.3.1 Electronic Component Boards Assemblies under
               Changes of Temperature 31
        3.3.2 Electronic Component Boards under Vibration and
               Mechanical Shock Loading 34
   3.4 Failures in Electronic Component Boards. 38
        3.4.1 Fracture Modes. 40
        3.4.2 Fatigue Failures 40
   References 42

4 Introduction to Thermodynamic-Kinetic Method 45
   4.1 Thermodynamics. 45
        4.1.1 Gibbs Free Energy and Thermodynamic Equilibrium . . . 46
        4.1.2 The Chemical Potential and Activity in a
               Binary Solid Solution 49

		4.1.3	Driving Force for Chemical Reaction	51
		4.1.4	The Phase Rule and Phase Diagrams	52
		4.1.5	Thermodynamics of Phase Diagrams	53
		4.1.6	Calculation of Phase Diagrams	66
		4.1.7	Local Equilibrium in Thin Film System?	71
		4.1.8	The Use of Gibbs Energy Diagrams	71
	4.2	Diffusion Kinetics		76
		4.2.1	Multiphase Diffusion	76
		4.2.2	Diffusion Couples and the Diffusion Path	80
	4.3	Microstructures		84
		4.3.1	The Role of Grain Boundaries	84
		4.3.2	The Role of Impurities	86
		4.3.3	Segregation	87
		4.3.4	Amorphous Structures	88
		4.3.5	The Role of Nucleation	89
		4.3.6	The Role of Steep Concentration Gradients	95
	4.4	Summary		97
	References			97
5	**Interfacial Adhesion in Polymer Systems**			101
	5.1	Adsorption		102
		5.1.1	Enthalpy of Adsorption	102
		5.1.2	Specificity	103
		5.1.3	Reversibility	103
		5.1.4	Thickness of the Adsorbed Layer	103
		5.1.5	Conditions of Adsorption	103
	5.2	Surface Energy and the Wetting of a Solid Polymer Surface		105
		5.2.1	Equation-of-State Model	107
		5.2.2	Surface Free Energy Component Models	109
	5.3	Kinetics of Wetting		112
		5.3.1	Role of Surface Roughness and Capillary Action	112
		5.3.2	Role of Chemical Surface Homogeneity	113
		5.3.3	Influence of Interphasial Interactions	114
		5.3.4	Influence of liquid viscosity	114
	5.4	Surface Modification		115
	5.5	A Brief Look at the Adhesion Mechanisms		118
		5.5.1	Adsorption Mechanism (Contact Adhesion)	118
		5.5.2	Diffusion Mechanism	119
		5.5.3	Mechanical Interlocking Mechanism	119
		5.5.4	Electrostatic Mechanism	119
		5.5.5	Adhesion Mechanisms in Practice	120
	5.6	Durability of Interfacial Adhesion		123

	5.7	Measurement of Interfacial Adhesion	124
	5.8	Work of Adhesion	126
		5.8.1 Testing the Adhesion Strength-Quantitative versus Qualitative	126
	References	128
6	**Evolution of Different Types of Interfacial Structures**	135
	6.1	Examples of Metal–Metal Interfaces	135
		6.1.1 Au/Al in Wire Bonding	135
		6.1.2 Sn-Based Solder Versus UBM in Flip Chip	140
		6.1.3 Ni or Ni(P) Metallisation on PWB Versus SnAgCu solder	151
		6.1.4 Redeposition of $AuSn_4$ on Top of Ni_3Sn_4	159
		6.1.5 Zn in Lead-Free Soldering	171
		6.1.6 Deformation Induced Interfacial Cracking of Sn-Rich Solder Interconnections	172
		6.1.7 Diffusion and Growth Mechanism of Nb_3Sn Superconductor	177
	6.2	Examples of Metal–Ceramic Interfaces	181
		6.2.1 Cu/diffusion Barrier/Si Systems	181
		6.2.2 Reactive Brazing of Ceramics	188
		6.2.3 Reactions Between Non-Nitride Forming Metals and Si_3N_4	192
	6.3	Example of Metallisation Adhesion on a Polymer Substrate	194
	6.4	Example of Polymer/Polymerisation Adhesion on Metal Substrate	198
	6.5	Example of Polymer Coating Adhesion on a Polymer Substrate	200
	6.6	Some Examples of Epoxy Polymer Surface Treatments	201
	References	204
Index		213

Chapter 1
Introduction: Away from Trial and Error Methods

The components and assemblies of microelectronics are complex multimaterial systems. They are typically composed of thin layers of metals, ceramics, polymers and semiconductors with numerous discontinuity areas, i.e. interfaces, between those distinctively dissimilar materials. The prevailing trends in microelectronics, namely increasing functionality and decreasing cost, has led to extremely complex multimaterial systems in ever smaller dimensions. New "More than Moore" type of approaches, the merge of ICs with sensors and transducers, 3D integration and new wafer level vertical interconnect technologies reflect the diversity of such systems. At the same time, the environmental requirements have increased. The requirements for electromobility adds to the harsh environment automotive electronics arena, more electrical functions are available in portable form and therefore easier to damage, implantable bioelectronics and sensors expose devices to interaction with human immune system and corrosive environments. Thus, there are three main facts which make the materials compatibility issues even more challenging in future: (i) number of discontinuity areas have increased, (ii) the smaller dimensions are more vulnerable to flaws and (iii) the loading conditions are more severe.

Which tools do the microelectronics and microsystems industries have available to tackle these increasing challenges and develop reliable systems for the new applications? Often both the frontend (FE) and backend (BE) development of microelectronics has, in the past, been forced to lean on extensive trial and error methods to search for the optimum selection of parameters between the design, bill of materials and processing for given loading conditions. Over several decades of research and development work much data has been collected and the accumulated knowhow on functional recipes for various microelectronics structures is impressive. But there are, however, numerous limitations related to this approach.

The accumulated knowhow and the functional recipes generated by purely empirical means are typically point solutions. So, if one considers the space between possible bill of materials (BOM), designs and load conditions (Fig. 1.1), several singular recipes with specific BOM, specific design and specific conditions

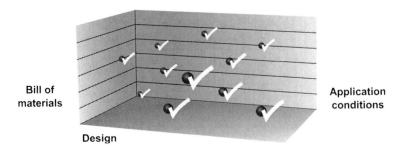

Fig. 1.1 Point solutions in a design of experiment landscape

are known to work. However, quite often it is not thoroughly understood why exactly this recipe works, and the role that different parameter changes play. This is crucial in a situation where changes in the system force to step out of this balance point. Due to continuous pressure to cut costs, on one hand, and development of new functionalities, on the other, this is in practice a typical situation. Only a slight change in the system, such as leaving out one metal layer from a metallisation schema, may destroy the alleged balance and new extensive experimentation is needed to reestablish system reliability. Thus, the complexity of nature surpasses our currently available functional recipes. Furthermore, the industry and also research and development work in this arena is lacking fundamental knowhow of the physics of failures as well as the interactions in the multimaterial systems.

Searching for optimum parameters via trial and error methods is also extremely time consuming. Often such methodologies will focus on only one problem at a time. It is, however, not uncommon that solving one problem locally gives rise to another problem in the vicinity. In this way, many hardware development teams unfortunately end up chasing the weak link in the system. When each issue will be addressed with individually tailored experiments and analysis, without revealing the real root cause for the initial issue, the overall progress becomes very slow and tedious.

Figure 1.2 illustrates a circulation of the weak link in an imaginary power package with a new solder die attach. In this case, it is possible to ensure that the die attach does not suffer from weak mechanical properties by first alloying the material with additives, which, however, leads to partial solder melting. This can be avoided by lowering the die attach temperature which will result in an incomplete interfacial reaction layer formation. This could be corrected by a change of backside metallisation, which, in turn, causes the formation of unfavourable intermetallics at the wafer backside/solder interface. Such examples can be found in all interconnect levels of electronics.

This book presents an alternative development methodology to the trial and error-based work, the target being to give basic knowledge and tools to solve complicated interfacial compatibility challenges in a wide variety of microelectronics and microsystems technology applications. The basic tools are drawn from the

1 Introduction: Away from Trial and Error Methods

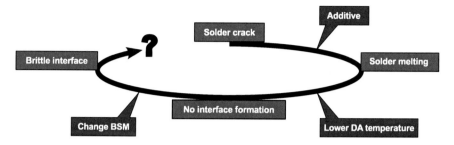

Fig. 1.2 Example of a circulation of the weak link in microelectronics development

materials science and engineering as well as from the mechanical engineering, and as such are not new. The novelty is in the combined usage of the methods and their application in electronics and microsystems technology. The book introduces a reader to the fields of thermodynamics and reaction kinetics of materials, theories of microstructures, mechanics of materials in multimaterial systems as well as to the mechanisms of adhesion.

The thermodynamics and reaction kinetics of materials, the theories of microstructures, and the mechanics of materials are the major constituents of materials science and there are plenty of high quality textbooks available focused solely on these specific areas. This book does not seek to compete with such reference books for teaching the theories of each scientific field. In this book, rather, these theories and background concepts are presented in the depth needed to combine the basic theories of materials science, and to solve multimaterials compatibility issues in microelectronics. The thermodynamics of materials has a central role in this book as it is such a powerful and generic method, and as there is nowadays an increasing amount of thermodynamic data available for practical applications.

As mentioned, the specific challenge of microelectronics in comparison to many other industrial applications is that the assemblies are typically composed of thin layers of numerous dissimilar materials and they can experience varying loading conditions. Hence, the main approach of this book is that four methods must be combined to understand and control the reliability of these complicated systems. These methods are (i) thermodynamics of materials (ii) reaction kinetics (iii) theory of microstructures and (iv) stress and strain analysis. This multifaceted approach is summarised in Fig. 1.3.

The thermodynamics gives us the criterion what can occur at the interfaces of the multimaterial system during processing or in service of a device. However, it does not contain any information about the time frame of the changes and thus must be combined with the reaction kinetics. If the necessary thermodynamic data of a system is combined with the available diffusion kinetics information and the detailed microstructural analyses, one can evaluate the stable and metastable phases of the system, their chemical compositions and relative amounts at different temperatures, as well as the evolution of phase structures in interconnection

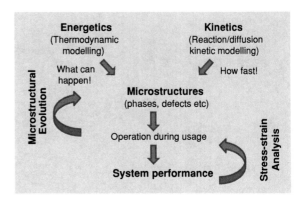

Fig. 1.3 Schematic presentation of the combined method

systems. With the help of a thermodynamic-kinetic method one can predict the sequence of phases (i.e. reaction products) in interconnections. What cannot, however, be predicted by combined thermodynamic-kinetic approach are the defect structures generated in various fabrication processes and in service of a product. Even though the formation and evolution of defect structures have been studied extensively during the last several decades, the theories are still mainly qualitative in nature, and therefore experimental research tools like optical, scanning and transmission electron microscopies have been employed to reveal the relations between microstructures and the mechanical and physical properties of materials.

When the results obtained with the above described methods are further combined with mechanical data and thermo-mechanical simulation, it is possible to understand the mechanical performance of an interconnection system, metallisation schema or another multimaterial structure. If we can predict the microstructure of an interconnection system, the external loadings concentrated on this material structure and how the multimaterial structure responds to those stresses, we can then conclude with a significantly reduced number of experiments whether the system is suited to the target specification.

It is important to note that interconnects and subsystems in microelectronics do not have a fixed phase and microstructure that determines the ultimate performance and the reliability of the system. Instead, the microstructures evolve continuously during the operation of a device—either slowly or more rapidly depending on internal and external loading conditions. Because the microstructures of interconnects and subsystems are thermodynamically unstable, they will change if the kinetics of a system allow this. However, since service temperatures in microelectronics and microsystems are relatively high for the materials typically used in the assemblies, the kinetics will allow new microstructures to form during the operation as it happens, for example, when solder interconnections will recrystallize locally or when an interfacial phase disappears or is replaced by another phase. What is more, the evolution of microstructures will change also the properties of the multimaterial systems as a function of time, which again has an impact on the performance of the system as a whole.

1 Introduction: Away from Trial and Error Methods

The book is organised as follows. Chapter 2 gives the reader an overview of the complexity and the diversity of multimaterial structures in modern microelectronics and microsystems. Chapter 3 presents how stresses are generated as a consequence of different material mechanical properties, and what is the response of multimaterial structures on those stresses. The thermodynamic-kinetic method is covered in Chap. 4, and the general principles determining the adhesion in typical polymer–polymer and polymer-metal systems are presented in Chap. 5. After reviewing the most important theories and tools, the new methodology which combines them is demonstrated in Chap. 6 with the help of several cases from the electronics industry. The examples include Cu impact on Au wirebonding, flip chip under bump metallisation reactions with solders, IC diffusion barrier reactions with Cu metallisation, SU-8 adhesion on Cu and many other multimaterial systems.

Chapter 2
Materials and Interfaces in Microsystems

2.1 Levels of Interconnections, Typical Stress Factors and Related Failure Mechanisms

The interconnections in electronic systems are traditionally divided into four (or more) categories: (i) interconnections on the IC-level, (ii) interconnections on the package level, (iii) interconnections on the board level and (iv) interconnections on the system level [1]. Since, the system level is product or application field-dependent, only the three levels, which are generic to all products, are considered here. Each level is characterised by typical materials and manufacturing processes. In addition, both the internal and external stresses vary between the levels. However, the ongoing trend to develop new solutions with higher integration has fused these interconnection levels closer to each other both dimensionally as well as technologically. Currently, materials and manufacturing methods that were previously used in IC (frontend) processes are utilised also in system level packages [2]. Since microsystems are heterogeneous structures (see Fig. 2.1) composed of many materials where components are integrated by different kinds of interfaces, the reaction between the materials and their environment decides manufacturability, functionality and reliability.

As can be seen from Fig. 2.1, within volume of less than 0.001 mm^3 there are materials from all basic material groups; polymers (epoxy and polyimide), insulators (SiO_2 and SiN), semiconductors (Si) and metals (Al, Cu, Ni, Sn and Ti) that all behave differently under varying temperature or stress states. Therefore, understanding on materials' compatibility is fundamentally important and provides the basis for comprehending the ever increasing electrical, thermal, thermo-mechanical and environmental challenges that must be faced.

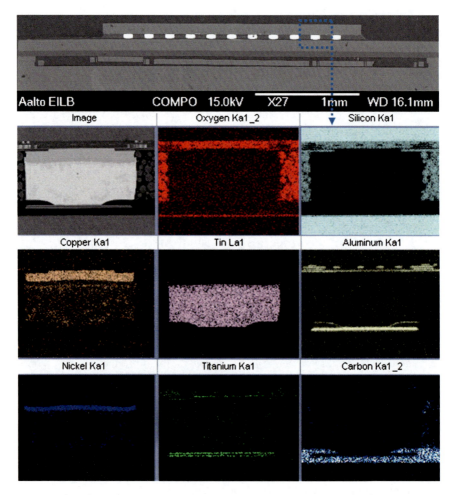

Fig. 2.1 SEM micrograph and EDS element maps from MEMS accelerometer with integrated ASIC

2.1.1 IC Level

Practically all microelectronic devices are based on "components" that are fabricated by employing metal-semiconductor contacts that are integrated with metallic connectors insulated by ceramic or polymer layers as shown in Fig. 2.2.

Semiconductor devices are based on a vast variety of thin conducting and insulating layers in contact with each other as well as to semiconducting materials [3]. Therefore, the interfacial compatibility between dissimilar materials is especially important on the IC-level. Typically, the main manufacturing and reliability challenges are related to the polycrystalline conducting thin films, which provide the electrical contacts between the devices and to the outside world via the packaging. It is to be noted that the physical and chemical properties of thin films

2.1 Levels of Interconnections, Typical Stress Factors and Related Failure Mechanisms

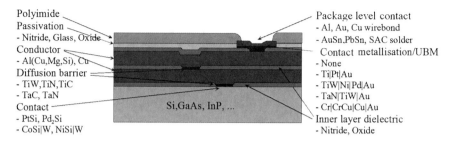

Fig. 2.2 Simplified schematic representation of typical frontside material solutions from an electronic component. (N.B. Layers are not in scale.)

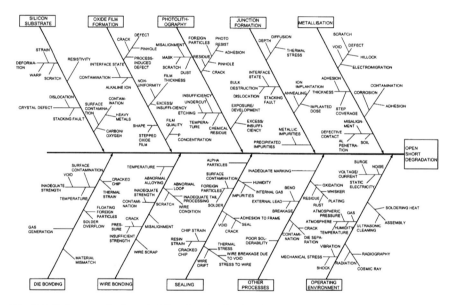

Fig. 2.3 Failure mechanisms in IC and packaging levels

can differ significantly from those of bulk materials. This, on the one hand, provides a huge amount of possibilities to tailor the material properties, but, on the other hand, makes the fundamental understanding of materials scientific aspects even more important, in order to avoid unwanted events. The key issue in comprehending the use of thin film structures is through understanding on diffusion and reaction mechanisms as well as the electrochemical and electromigration properties of these materials. The huge range of different factors affecting the failure mechanisms on the IC and packaging level is illustrated in Fig. 2.3 by the fish-bone diagram.

The above described aspects are mainly related to "traditional" ICs and the relative importance of these aspects can be quite different when one considers for instance high power or RF-components as well as LED or MEMS systems. For

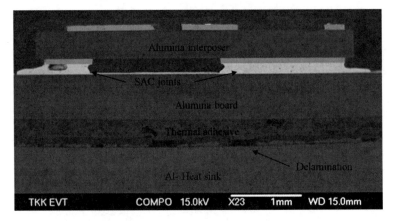

Fig. 2.4 Heat-sink attachment failure in high-power LED product after thermal cycling 0 ⇔100°C, 2 h cycle time and 500 cycles (N.B. The LED component is not seen in this section)

example, the major challenges in RF-circuits and modules, in addition to increasing power consumption and difficulties in thermal management, are related to low noise and high linearity issues. Therefore different approaches must be taken for both manufacturing methods and materials selection points of view. On the other hand, the performance and reliability of LED components are greatly dependent on thermal management since high operating temperatures at the LED junction adversely affect performance, resulting both decreased output and lifetime. It is to be emphasised that the majority of LED failure mechanisms are temperature-dependent. When designing lighting systems using high-power LEDs, the most important guideline is to minimise the amount of heat that needs to be removed and enhance the thermal conductivity between the heat sink and the LED. Since the LED components are typically very reliable, it has been shown that many lighting applications the packaging or, especially, board-level interconnection reliability defines the lifetime of the product. Figure 2.4 shows a cross section from a high-power led-module after 500 cycles of thermal cycling, where the thermal adhesive between the alumina board and aluminium heat sink has failed. In addition, it is important to separate the LED drive circuitry from the LED board so that the heat generated by the driver will not contribute to the LED junction temperature.

2.1.2 Package Level

As was already mentioned above, the integration of microsystems is continuously blurring the distinction between the interconnection levels. However, since on the packaging level the volume drivers have continuously shifted towards consumer electronics, many material and manufacturing solutions, like ceramic packaging,

2.1 Levels of Interconnections, Typical Stress Factors and Related Failure Mechanisms

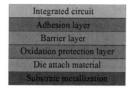

Fig. 2.5 Schematic presentation of typical die attach structure and commonly used materials

Integrated circuit	Silicon, GaAs,...
Adhesion layer	Ti,TiW,Cr,...
Barrier layer	Ni, NiV,Ni(P), Co, Co(P), NiCo,...
Oxidation protection layer	Au, Pt,...
Die attach material	AuSn20, PbSn10, Ag-epoxy, Ag-glass,...
Substrate metallization	ENIG, ENEPIG, Cu|OSP, Cu|Sn, CuAg,...

must be ruled out due to cost factors. Therefore, in this "cost-performance" category the methods used and developed are based on the low-cost polymer and solder materials and technologies.

Die attach (also known as die mount or die bond) is the process where the silicon chips are mounted on the substrate. The most commonly used methods are adhesive or glass bonding and soldering or eutectic die attach. In addition to mechanical support, the die attach materials also provide thermal and/or electrical conductivity between the die and the package thus affecting the performance of the device during field operation. When solder materials are used in the die attach, backside metallisation of the semiconductor device is required. The backside metallisation is typically a multilayer structure composed of an adhesion layer, a barrier/wetting layer and an oxidation protection layer (See Fig. 2.5).

The requirements of a die attach material are application-dependent, but typically include that (i) mechanical strength must be maintained up to 150°C, (ii) the attachment process temperature must not affect the die function and (iii) the attachment should withstand board-level assembly processes (temperatures up to 260°C for a few minutes), the absorption of continuous and cyclic mechanical and thermomechanical stresses, etc. The most common stress-related failures during product life are die/adhesive cracks, passivation cracks, pad delamination and corrosion [4]. Figure 2.6 shows an example of a cracked Si-chip due to lack of mechanical support of the die attach film during wire bonding in a stacked die application.

Wirebonding has been the most commonly used method for connecting ICs to packages or other substrates already now for many decades. The most important advantages are related to the flexibility of the manufacturing process, long experience and reliability. Wirebonding can also be considered as low-cost from an equipment and materials used point of view. There are three methods used: Thermocompression (T/C), Thermosonic (T/S) and Ultrasonic (U/S) (see Table 2.1). Depending on the method, the shape of the bond can be either "ball-wedge" (T/C- or T/S-methods) or "wedge–wedge" (U/S-method). Aluminium (Si or Ge alloying) wires are typically used with the U/S-method. T/C- and T/S- methods typically use Au-wires, but Pd, Cu and Ag are also possible.

The challenges in wirebonding are frequently related either to bonding process parameter control (temperature, pressure, etc.) or to reliability problems caused by high operational temperatures or aggressive environments. Figure 2.7 shows optical micrographs from a typical process-related problem, where the Al pad has been consumed locally by the IMC-reaction with Au-ball due to excessive bonding temperature.

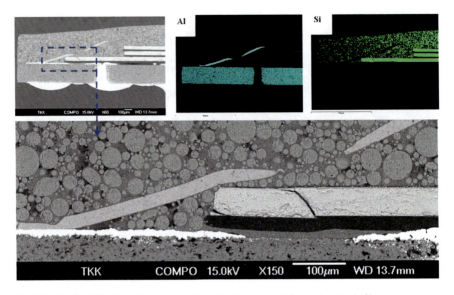

Fig. 2.6 Cracked Si-chip due to lack of mechanical support from die attach film

Table 2.1 Three wirebonding methods and their typical properties

	Thermocompression (T/C)	Thermosonic (T/S)	Ultrasonic (U/S)
Definition	Contact between wire and contact pads is based on plastic deformation and interfaces	Ultrasound is used to assist the disruption of the oxide layers	Ultrasound is used together with pressure but no extern al heating
Temperature	300–400°C	150–200°C	Room temperature Bond interface ∼7 70–80°C
Pressure	100 g/contact area		
Time	0.1 s	5 ms < t < 25 ms	t < 25 ms
Bond type	"ball-wedge"	"ball-wedge"	"wedge–wedge"
	Factors affecting reliability: Surf ace roughness and voids, oxidation. adsorbed impurities, moisture	Used especially when wire bonding thick-film hybrid substrates	Problems in "wedge–wedge" is that both contacts have to be aligned in the same direction
Wire	Usually Au	Au. also Cu and pd	Al

Two typical operational stress-related reliability problems are shown in Figs. 2.8 and 2.9. Local corrosion of an Al pad in a Cl^- containing environment can be clearly detected from Fig. 2.8. The EDS element maps show higher chlorine and oxygen content at the surface pits. Figure 2.9 shows the well-known "purple plague" problem, which is the formation of an excessive amount of $AuAl_2$ intermetallic at the interface [5]. The growth of AlAu intermetallics is often accompanied by the formation of so-called Kirkendall voids, which are caused by

2.1 Levels of Interconnections, Typical Stress Factors and Related Failure Mechanisms   13

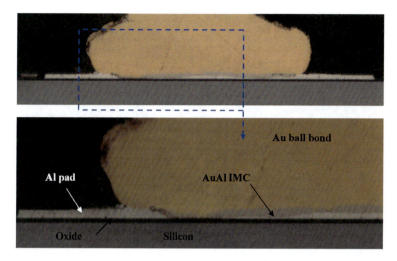

Fig. 2.7 Au ball | Al pad interfacial reaction during the bonding process

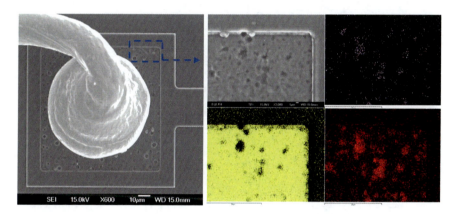

Fig. 2.8 SEM micrograph from local corrosion at the Al-bond pad together with EDS element maps (violet_chlorine, yellow_aluminum, and red_oxygen)

the differences between the intrinsic diffusion fluxes of Al and Au, i.e. Au moves faster than Al and thus the vacancies accumulate on the Au side.

Tape Automated Bonding (TAB) technology was developed as a low-cost method during the 1960s for small sale integration (SSI) circuits to replace wirebonding. During the 1980s, the use of surface mount technology made it more popular also in high I/O components (see Fig. 2.10). The TAB process uses typically bumped chips, which are first bonded to metallised tape fingers (ILB, Inner Lead Bonding). After ILB the chips are tested and/or encapsulated before singulation and outer lead bonding to a substrate or card.

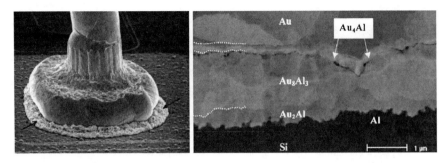

Fig. 2.9 *Left* "purple plague" and *right* AuAl IMC layers [5]

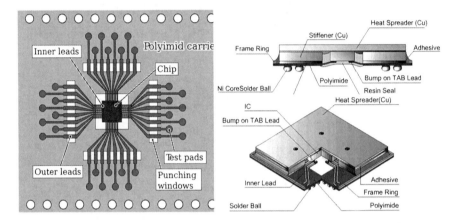

Fig. 2.10 TAB used in a component package

Flip-chip is defined as a bare chip mounted on the substrate upside down, the active side towards the substrate. Flip-chip technology is the so-called Area Array method, which means that the whole IC area can be used for interconnections. Therefore flip-chip provides higher interconnection densities than both wire bonding and TAB. The I/O contact pads are located under the chip, and connections to the substrate can be made with various interconnection materials and methods (see Fig. 2.11), such as solder bumps, conductive adhesives or Stud Bump Bonding (SBB). The substrate can either be a component (like CSP or BGA) interposer or a mere PWB (Flip-Chip On Board, FCOB), ceramic (FCOC), glass (FCOG) or flex (FCOF) substrate.

From an electrical performance point of view flip-chip is also a superior direct chip attachment method. Due to the advantages of the technology it is commonly used in microprocessor components. However, as a result of the small height of the solder interconnections, which are the most commonly used, the large difference in the coefficients of thermal expansion (CTE) between the Si-chip and the organic substrate cause reliability problems. Large strains are already formed during the

2.1 Levels of Interconnections, Typical Stress Factors and Related Failure Mechanisms 15

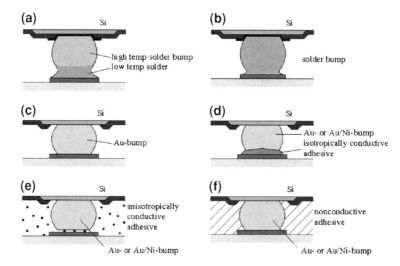

Fig. 2.11 Comparison between different flip-chip connection types [6]

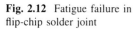

Fig. 2.12 Fatigue failure in flip-chip solder joint

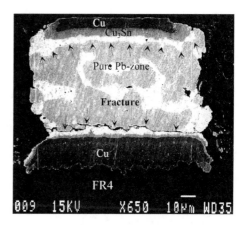

cooling in the soldering process and changes in operational temperature lead to thermomechanical low-cycle fatigue. Figure 2.12 shows a fatigue failure from a flip-chip interconnection.

Another challenge related to solder interconnections in flip-chip joints is caused by the small solder volume. During the soldering process (and consequent high-temperature use) a relatively large fraction of the solder material can be consumed by the interfacial intermetallic reactions with component and/or PWB metallisations. A particularly challenging situation is faced if only one element (usually Sn) is consumed, but the other element (like Pb or Bi) takes no part in the chemical reactions. This leads to local enrichment of the non-reactive element near the interface, which can thus become very brittle. This small volume effect can also be seen in Fig. 2.12, where the fatigued crack clearly follows the interface between the IMC-layer and Pb- enriched area. [7]

Table 2.2 Properties of packaging materials [8]

Material (wt%)	Melting point/ Softening temp. (°C)	CTE (ppm/K)	Thermal conductivity (W/mK)	Electrical resistivity (Ωm)	Young's modulus (GPa)
Alumina	~2,000	6.5	37	>10^{14}	370–410
AlN	~2,800	5.7	140–220	>10^{13}	350
SiC	~2,500	3.7	270	>10^{13}	420
Glass	Soft >450	5.3	0.4–1.3	7.5×10^{11}	
Alloy 42	1,425	4.5–5.5	14.6	72×10^{-8}	148
Nickel	1,455	12.7	91	6.8×10^{-8}	214
Copper	1,084	16.8	386	1.7×10^{-8}	115–130
Au80Sn20	280	16	57	16.4×10^{-8}	59
Au97Si3	370	10–13	94		83
Pb95Sn5	305–315	29	32.3		7.4
Ag epoxy	Soft. >300	30–55(below Tg)	0.8	1.4×10^{-6}	
Sn95.8Ag3.5CuO.7	217			13×10^{-8}	56

Moulding/Encapsulation/Underfilling is typically the final manufacturing step in a non-hermetic packaging process. Hermetic packaging requires the use of ceramic, glass or metallic materials that makes this approach too expensive in consumer electronic products. Plastic encapsulation also provides lighter and smaller packages compared to ceramic counterparts. The major difficulties are related to the process, which is still not well characterised or optimised and reliability. The complexity of the rheokinetics of thermosetting encapsulants, the complex mould geometries and conditions can lead to incomplete filling of the mould, void formation and transport, wire sweep and die pad shift.

The main purpose in both (hermetic and non-hermetic) packaging methods is to provide protection to circuits from damage during next level assembly the service life environmental stresses. In addition, the package provides a path for the thermal energy to be dissipated from the IC. Table 2.2 shows typical physical, mechanical and electrical properties of packaging materials

Wafer Level Packaging (WLP) defined as a technology, where packaging and interconnections are fabricated on the wafer prior to dicing and no other packaging steps are performed prior to board assembly. The process eliminates a few conventional packaging steps like die attach and wire bonding and thus provides batch processing to lower costs, streamlining of handling and shipping logistics, minimal form factor and the avoidance of known good die (KGD). It also facilitates faster time-to-market. Furthermore, when wider pitch is achieved it allows for wider UBM, taller bumps, and thus better joint reliability. The major challenges of WLP are related to (i) an infrastructure that is not fully established, (ii) the high cost for poor yield wafers and wafer bumping, (iii) die shrink strategies and, (iv) in particular, solder joint reliability (since underfilling cannot be used in all applications). The development of so-called pre-applied or wafer level underfills may provide solution for this in future. Another important question is whether the

2.1 Levels of Interconnections, Typical Stress Factors and Related Failure Mechanisms 17

Fig. 2.13 Wafer bonding techniques

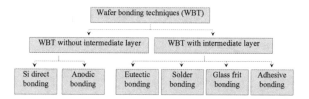

package manufacturer is an IC or bumping house. Currently WLP applications are typically passives and discretes as well as RF- and memory components where the I/O amount is up to a few hundreds, but the WLP approach is rapidly expanding to logic ICs and MEMS components (already shown in Fig. 2.1).

Wafer bonding techniques have recently become more attractive in wafer level packaging of MEMS components since they are able to meet the versatile requirements placed upon them: like for instance, a vacuum inside an hermetically sealed package or simultaneous protection from the environment while at the same time alloying chemical, pressure, flow, optical or other measurement, detection or actuation. Wafer bonding (see Fig. 2.13 and Table 2.3) includes direct bonding methods like anodic and fusion bonding and methods that utilise intermediate metal or insulator layers. Generally, it can be stated that wafer bonding is challenging since the thermal expansion of the wafers produce alignment problems. An even bond over the entire bond-surface, and clean bond surfaces are required.

Anodic (or electrostatic) bonding utilises temperatures of up to 550°C and high voltages of up to 2,000 V to obtain a chemical process, which creates a permanent joint between the silicon and sodium rich glass. This way an insulating, biocompatible and highly hermetic package without CTE problems is achieved. However, due to the relatively thick glass layers needed it is not optimal for thin-film packaging. In addition, surface roughness better than 500 nm is required. Fusion bonding is based on formation of chemical Si–OH (hydrophilic) bonds or Si–H (hydrophobic) bonds. Fusion bonding requires extremely smooth (<50 nm), clean and flat wafers and utilises temperatures up to 1,100°C. The main advantages of fusion bonding are strong connection due to covalent forces, relatively free choice of wafer material, frontend compatibility, fast process and wide process window. The main challenges are related to cleanliness and surface quality requirements and that the standard annealing process is not CMOS compatible.

Metal bonding methods, which are most common in MEMS packaging, include solder bonding, eutectic bonding, thermocompression bonding (TCB) and rapid thermal processing (RTP). They are used to seal "rough" surfaces at low temperatures. In eutectic bonding two thin films on both wafers form an alloy. This typically is achieved while one thin film (pure metal) in contact with the alloy on the second wafer is heated above the melting point of the alloy. Alignment, temperature and bond pressure need proper control. Also electrical feedthroughs may cause problems and wafers are limited in the topography. In glass frit bonding the printing and alignment are required before the wafers are bonded using a combination of pressure and heat. The glass frames serve as an adhesive and

Table 2.3 Wafer bonding techniques

Method	Material	Temperature (°C)	Comments
Anodic	Glass to Si-glass	180–550	Small amount of surface roughness hermetic
Fusion	Si to Si	>800	Very small surface roughness. hermetic
Eutectic	Au–Si, AI–Ge, Au–Sn	Typically 260–420 system-dependent	"Rough" surfaces, hermetic
Solder	AuSn, PbSn, InSn or AISi to Si and glass	System-dependent from 118–800	"Rough" surfaces. hermetic
Thermocompression bonding (ICB)	Au or Cu to Si	25–250	High force required hermetic
Rapid thermal processing (RIP)	Al to Si_3N_4	750 Short duration	Hermetic
Adhesive	SU-8, BCE, PI	<300	"Rough" surfaces, non-hermetic
Glass	Reflowed glass frit	375–410	"Rough" surfaces hermetic

sealant between the wafers. The main advantages are related to a very flexible and economical process, a relatively free choice of wafer material, the fact that the bond surfaces do not need to be very smooth and that the glass adapts itself to the wafer topography. The main disadvantages are related to the relatively "dirty" process and small process window.

Printed circuit/wiring boards (PCBs/PWBs) and interposers can also be considered to belong to the package level. The miniaturisation, environmental aspects and increased performance of the modern packages have forced the evolution of printed circuits too. The current requirements for PCBs include high density wiring, bare chip attachment capability, smallness of size, a light and thin form factor, high reliability, cost efficiency, functionality at high frequencies, high thermal conductivity and an environmentally friendly manufacturing process and recycling. Generally PCBs have three main functions: (i) To support the mechanical structure of the device, (ii) To be a foundation for electrical connections between components and (iii) To help in removing thermal power. Since the PCB is one of the most expensive components in an electronic device, it is necessary to ensure the manufacturing quality of this component. On the other hand the reparability of a PCB after assembly is usually impossible or at least much more difficult than repairing (changing) an individual component/solder joint. It is also to be emphasised that there are not "standard" PCBs for multiple customers and products but each board represents an application-specific design. PCBs can be classified with different means for instance according to manufacturing process or based on the dielectric or core material. As seen from Fig. 2.14 in the most commonly used classification the metal core printed circuit boards (MCPCB), which are typically used in thermally demanding applications like LED modules, are included under rigid organic boards. [9]

Fig. 2.14 Classification of printed wiring boards

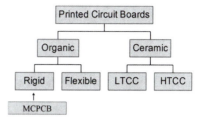

Fig. 2.15 iSiP structure with one embedded component. Substrate top layer is utilised by one wire-bonded die, moulded (Courtesy of Imbera Electronics)

The advances in substrate technology play a crucial role in achieving the aim of microminiaturised multifunctional systems. Traditionally, the substrates have played a passive role, but currently the embedding of active components in PCBs is becoming a significant option.

The embedded components are currently used in various application areas, today mainly hand-held consumer electronics since that can provide miniaturisation/form factor reduction ($x-y$, and thickness) simultaneously with excellent electrical and thermal performance. Embedding techniques for ceramic substrates were developed already during the 1980s, but due to high material and process costs their applicability was limited to only a few high performance devices [10, 11]. During the 1990s the emphasis was placed on utilising organic substrates and known PCB manufacturing processes like laser drilling, photolithography, etching techniques and chemical or electrochemical copper deposition [12–14]. The current trend is to implement embedding techniques also in System-In-Package modules (see Fig. 2.15) utilising novel 3D packages with z-axis connections. This enables the usage of the package top surface for package stacking

Embedding can include active integrated circuits or passive components such as resistors, capacitors or integrated passive devices. Typically, a terminal metallurgy—for example 5 μm copper—is required on embedded component pads. Figure 2.16 shows an example of a typical process flow and die level interconnection made with the IMB® technique.

The utilisation of component embedding techniques, especially at the component level, can offer a highly accurate and cost effective way of making interconnections. These benefits can be achieved with a wide variety of embedded components by utilising panel-sized volume production. Since the interconnections can be considered as pure Cu–Cu the long-term reliability is inherently

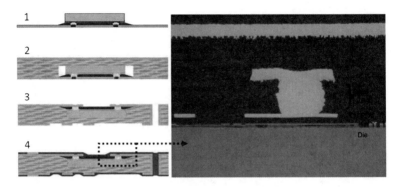

Fig. 2.16 Left_IMB® 3.0 generation process flow Right_Die copper terminal (bump/RDL) is connected to PCB copper layer with IMB microvia (Courtesy of Imbera Electronics)

Table 2.4 Commonly used solders, PCB and component metallisations [15, 16]

Solder (wt%)	PWB metallisation	Component metallisation
Sn–Pb	HASL (SnPb, SnCu)	Aluminium
Sn–Bi	Cu|OSP	Bright tin
Sn–Sb	Electroless Ni Immersion gold	Chromium
Sn–Ag	Electrolytic Ni Electroplated Au	Gold
Sn–In	ENEPIG	Indium
Sn–Zn	Cu|Immersion Sn	Matte Sn
Sn–Aez	Cu|Immersion Ag	Ni|Au
Sn–Cu–Ni		Ni|Pd|Au
Sn–Ag–Bi		Cu|OSP
Sn–Ag–In		Palladium
Sn–Ag–Zn		Pt|Pd|Ag
Sn–Ag–Cu–Ni		Silver
Sn–Ag–Cu–Sb		Ag|Pd
Sn–Ag–Bi–Cu–Ge		Tin and Sn based alloys
Sn–Ag–Cu–RE		Zink

excellent. The challenges with embedding techniques are related to requirements of high level material and process quality control required to achieve a high yield. Moreover, since no rework is possible with embedded components, KGD is required to achieve cost efficiency.

2.1.3 Board Level

The interconnections at the board level are typically made by soldering. During the last decade the soldering has changed markedly due to both environmental as well as manufacturability reasons. For more than five decades the components were

2.1 Levels of Interconnections, Typical Stress Factors and Related Failure Mechanisms   21

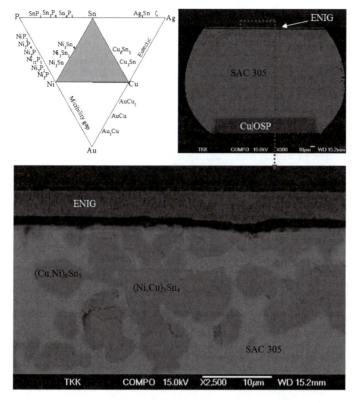

Fig. 2.17 **a** Schematic representation of a six component thermodynamic description including the binary IMCs, **b** SEM micrograph from SAC305 solder interconnection and **c** solder microstructure from SAC305 | ENIG interface

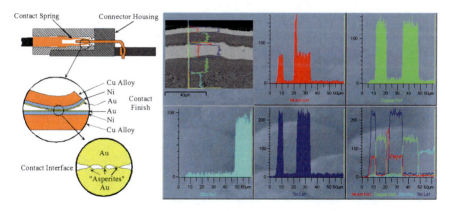

Fig. 2.18 Left_ Schematic illustration of a connector cross section (redrawn from 17). Right_ Example of contact finish layers from a DIN connector

attached to boards with SnPb solder, but the prohibition of lead changed the situation drastically. Even though, the soldering is still grounding on the chemical interaction between tin and typically copper or nickel contact metalisation, the implementation of lead-free solders and many different types of PCB or component metallisations has made the metallurgical system very complex. Table 2.4 shows commonly used solders, PCB and component metallisations.

Since the amount of different combinations is still increasing inevitably as, for example, new alloying elements to SAC solders are implemented, the manufacturability and reliability analyses purely by means of trial and error becomes impractical. For example, if one wishes to analyse the microstructural evolution of solder interconnections (see Fig. 2.17) at a constant temperature, where the contact metallisation is electroless nickel and immersion gold (ENIG) on the component side, Cu|OSp on board side and the solder is SAC305 (Sn 3.0Ag 0.5 Cu), a six component thermodynamic description (over 20 phases, even excluding ternary and higher order phases) is required.

The connections between circuit boards or products are characterised by utilisation of cables and connectors. The functionality and reliability of electromechanical connections has been the key enabling factor in system-level performance for decades. However, currently—especially in high-speed connections—optical signal transmission is utilised. Even though the solutions may be totally different between optical and electrical connections, the majority of the challenges are still related to (contact) interfaces (see Fig. 2.18) [17]. In electromechanical connectors the functionality and its stability over time is based on the interfacial phenomena at the contact surfaces. Typically, only a small fraction (even less than 1%) of the whole contact area is carrying the current through s.c. asperites or a-spots and, therefore, it is necessary to have a thorough understanding of the effects of parameters like, adsorption contamination, wear, fatigue, stress relaxation, etc.

There remain still very many challenges related to optical interconnects since the contamination of contact surfaces can cause severe power losses. Fluxless soldering is one of the possibilities in board-level interconnections, but a great many new innovations are still required before optical signal transmission can be adopted from system/product level down to component board or component level. The advanced SOP approaches may provide solutions in the future also for mass production. [2]

References

1. W. Brown, *Advanced Electronic Packaging—with emphasis on Multichip Modules* (IEEE Press, New York, 1999)
2. R. Tummala, *Introduction to System-on-Package—Miniaturization of the Entire System* (McGraw-Hill, New York, 2008)
3. J. Liu, O. Salmela, J. Särkkä, J.E. Morris, P.-E. Tegehall, C. Andersson, *Reliability of Microtechnology* (Springer, New York, 2011)

References

4. J.H. Lau, *Thermal Stress and Strain in Microelectronics Packaging* (Van Nostrand Reinhold, New York, 1993)
5. N. Noolu, N. Murdeshwar, K. Ely, J. Lippold, W. Baeslack, Phase transformations in thermally exposed Au–Al ball bonds. J. Electron. Mater. **33**(4), 340–352 (2004)
6. IVF, The Nordic Electronics Packaging Guideline, Chapter: B. Flip-Chip
7. K.J. Rönkä, F.J.J. van Loo, J.K. Kivilahti, The local nominal composition–Useful concept for microjoining and interconnection applications. Scripta Mater. **37**(10), 1575–1581 (1997)
8. J.E. Sergent, A. Krum, *Thermal Management Handbook for Electronic Assemblies* (McGraw-Hill, New York, 1998)
9. C. Coombs, *Printed Circuits Handbook* (McGraw-Hill, New York, 2001)
10. J.F. Burgess, C.A. Neugebauer, RwoH integral package for MCM using the GE HDI process with a metal barrier, *Electronic Components and Technology Conference, in Proceedings*, (1993) pp. 948–950
11. W. Daum, W. Burdick, R. Fillion, Overlay high-density interconnect: a chips-first multichip module technology. Computer **26**(11), 23–29 (1993)
12. A. Kujala, R. Tuominen, J.K. Kivilahti, Solderless interconnection and packaging technique for active components, in *The Proceedings of the 49th IEEE Electronic Components and Technology Conference*, San Diego, 1–4 June 1999 pp. 155–159
13. R. Tuominen, J.K. Kivilahti, A novel IMB technology for integrating active and passive components, in *The Proceedings of The 4th International Conference on Adhesive Joining & Coating Technology in Electronics Manufacturing*, Helsinki, 18–21 June 2000, pp. 269–273
14. A. Ostmann, A. Neumann, Chip in Polymer–Next Step in Miniaturization, Adv. Microelectron. **29**, 3 (2002)
15. IPC 7095B, Design and Assembly Process Implementation for BGAs. Accessed March 2008
16. K. Puttliz, K. Stalter, *Handbook of Lead-Free Solder Technology for Microelectronic Assemblies* (Marcel Dekker, New York, 2004)
17. R. Mroczkowski, *Electronic Connector Handbook* (McGraw Hill, New York, 1998)

Chapter 3
Introduction to Mechanics of Materials

Solid materials are commonly grouped into four basic categories based on their chemical nature and physical (atomic) structure: metals, polymers, ceramics and semiconductors. In addition to these there are the combinations of materials in the first three groups, namely composites. In electronic assemblies, various materials from all of these groups are brought together. Metals are used, for example, to form electrical connections in the form of copper traces on printed wiring boards (PWB) and wire bonds. They are also used as coatings to protect surfaces from oxidation in order to preserve the solderability of component I/Os and PWBs or to preserve good electrical contact between the surfaces of connectors. Polymers and elastomers are used as packaging and encapsulation materials, whereas polymers and ceramics are used as the substrates of the component assemblies, modules or individual components. Ceramics are used mainly because of their dielectricity and good thermal insulation. In addition, they are quite resistant to even the highest temperatures in electronics assembly processes. The FR4 type PWBs are composed of woven glass fibres impregnated with resin and copper layers, and they are one example of a much used composite material in electronic applications.

The employment of new materials in electronic devices and assemblies has brought additional challenges to design and production of electronic products. Over the past few years many commonly used materials have been changed to more environmentally friendly or less expensive alternatives. For example, halides have been removed from the printed wiring boards, and lead has been removed from solders and protective coatings. Copper is also expected to replace gold as wire bonding material in most applications in the near future. Furthermore, improvement in the physical properties of materials is also a strong driver for materials development. The introduction of nanoscale fillers into conventional polymers and composites has provided new opportunities to develop materials with improved physical properties, and even multifunctional behaviour. However, the adoption of new materials has given rise to a variety of reliability concerns which are mainly related to compatibility of new materials with the existing materials and processes.

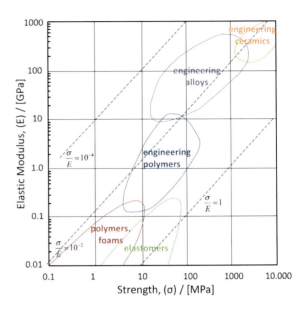

Fig. 3.1 Materials in electronic assemblies with respect to their strength and elastic modulus (adapted from [1])

It is particularly challenging to design reliable electronic assemblies since the commonly used materials are very different in their mechanical properties. Materials are different, for example, in their elastic modulus, plastic behaviour and coefficient of thermal expansion that are all essential material characteristics for overall product reliability. Figure 3.1 shows a map where some commonly used materials in electronic devices are placed with respect to their mechanical properties, namely strength and elastic modulus. Materials at opposite edges of this two-dimensional space are often used in contact with each other to form complex three-dimensional structures that are exposed to variety of different loadings in their use environments. In addition to the differences in mechanical properties compatibility between different materials has to be taken into account in the design of electronic products. Compatibility is particularly important in the joining of metals because the microstructures of materials ultimately determine the reliability of electronic assemblies, the field performance of each material combination, however, is likely to be different. The fact that ever more complex electronic assemblies are composed of many dissimilar materials emphasises that a good understanding of their properties is essential in order to design reliable electronic products. Making decision regarding choice of materials is not easy since they should not be made based on only a few decision criteria.

In the following we first examine briefly the mechanical properties under the assumption of uniaxial and uniformly distributed stress. However, this is often too simplistic and is especially so in the case of electronic component boards in operation conditions. Therefore it is important to bear in mind that the one-dimensional examination should be extended to three dimensions. However, this goes beyond the scope of this chapter. The formation of strains and stresses will be discussed next and, finally, the response of electronic component boards to different loading types, namely cyclic thermomechanical and mechanical shock loading, will be reviewed.

3.1 Deformation of Electronic Materials

Materials commonly used in electronic devices vary in their mechanical properties (see Fig. 3.1). Elastic deformation (i.e. deformation recoverable upon release of stress) in most metals, solders in particular, is very small as measured by highly sensitive equipment. For structural metals the elastic region extends up to about 0.2% but for common solders the region is much smaller [2, 3]. At stress levels equal to or higher than the yield stress, deformation is not recoverable upon release of stress and the material is deformed plastically. The stress–strain relation in the elastic region is commonly known as Hooke's Law ($\sigma = E\varepsilon$, where σ is stress, E is elastic modulus and ε is strain), which is the most well-known constitutive relation in continuum mechanics. In simple terms Hooke's Law states that strain is directly proportional to the stress. The generalised Hooke's Law can be used to predict the elastic deformations of a material by an arbitrary combination of stresses. The generalised Hookes Law shows how under three-dimensional loading all six stress components are linearly related to the six strain components through the Poisson ratio (the ratio of relative contraction to relative stretching). The isotropic form of Hooke's Law is employed most often but, in the case of certain materials used in electronics, this is not sufficient since they are either single crystals (e.g. Si) or they exhibit significantly anisotropic mechanical properties due to their asymmetry of crystal structure (e.g. Sn).

When loading is carried out above the yield stress, the load typically has to be increased for additional strain to occur. Figure 3.2 shows some examples of the

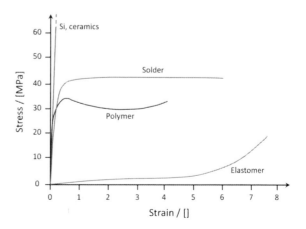

Fig. 3.2 Stress–strain behaviours of some electronic materials

stress–strain relations of materials commonly used in electronics. Ceramics and silicon are different from other material in electronics in the sense that they are very stiff (high elastic modulus) and their yield strength is relatively high but they often break before stresses can reach the yield stress. They are very brittle and they do not exhibit plastic behaviour. The stiffness and strengths of ceramics are comparable to those of the metals. However, nearly all other metals in electronic devices apart from Si exhibit plastic flow and strain hardening, which is associated with increase in the number of defects caused by plastic deformation. In strain hardening the strength of the material gradually increases until necking begins and the sample fails (the case of uniaxial tension). Polymers exhibit similar plastic deformation as the polymeric chains orient themselves in the direction of tension until necking begins in a way similar to metallic samples. However, polymers exhibit further strain hardening after the necking is initiated as the polymeric chains elongate, rotate, slide and detangle in the necking part of the specimen.

Currently there is no theoretical way to predict—based on uniaxial loading—the conditions at which plastic yielding begins when a material is subjected to any combination of stresses in a multiaxial state. The presently employed ways to predict yielding in multiaxial state are mainly empirical and there are two generally accepted approaches: (i) the distortion-energy criterion (von Mises: yielding occurs when the energy of distortion reaches the same energy for yield as under uniaxial stress) and (ii) the maximum-shear-stress criterion (Tresca: plastic flow of a metal begins when the maximum shear stress reaches the shear yield strength of under uniaxial load). The distortion-energy criterion is used in some of the commercial finite element analysis software packages to predict yielding in component boards. The theories of plastic deformation are often based on the assumption that the volume of solid remains constant during deformation whereas Hooke's Law takes small changes in volume into account with the help of Poisson's ration. Readers interested in a more thorough discussion of yielding under multiaxial loading can, for instance, turn to [4, 5].

It should be noted that the mechanical properties of all materials are dependent on temperature. With an increase in temperature the ability to carry loads is typically decreased. With increasing temperature (i) the value of elastic modulus decreases, (ii) the values of yield strength and ultimate tensile strength decrease, and (iii) ductility generally increase (there are few exceptions, however) unaffected, and (iv) strain hardening is less significant. Furthermore, the classification of deformation as elastic and plastic is valid only at relatively low temperatures. At the glass transition temperature region, polymers undergo a solid form transition from a glassy amorphous to viscous amorphous state and this transition is associated with a notable change in mechanical properties. The glass transition temperature range is also sensitive to humidity.

The mechanical properties of metals change markedly as temperature is increased above about 0.3 to 0.4 in the scale of homologous temperatures (the homologous temperature is defined as the ratio of the prevailing temperature to the melting point of a material expressed in absolute temperature scale). Namely, at high homologous temperatures the plastic deformation metals become time-dependent. There are two practically significant consequences: Firstly, plastic behaviour becomes strain-rate dependent: in general, strength increases with strain rate. Very often, the ductility is reduced with an increased rate of strain. This behavior is due to the limited ability of the dislocation-based deformation mechanisms to transmit instantaneous plastic deformation. The tendency to exhibit strain rate hardening depends, among other things, on the crystal structure and average velocity of dislocations in a material [6, 7]. Propensity towards twinning has been found to increase with higher strain rates because it can provide an additional mechanism of plastic deformation under high strain rates. Secondly, at high temperatures the time-dependent deformation mechanisms become faster and the division of plastic deformation into elastic and plastic is not unambiguous as metals can be deformed plastically even at stress levels below their macroscopic yield stress. When a load is applied over a long period of time, plastic strain increases with time, i.e. the material creeps, and the total strain is the sum of the three parts: (1) elastic strain, (2) time-independent plastic strain and (3) time-dependent plastic strain (i.e. creep). The failure criterion under creep conditions is complicated by the fact that stress is not proportional to strain. Readers interested in a more thorough treatment of the deformation mechanisms are directed to, for instance [8, 9].

The mechanism of time-dependent plastic deformation has great practical significance especially in the case of most solders used in electronics since room temperature is relatively high with respect to their melting temperatures. For example, room temperature of 23°C for eutectic SnAgCu solder alloy is about 0.6 in terms of homologous temperature. Furthermore, many of the mechanical properties of tin are anisotropic due to the asymmetric body centric tetragonal crystal structure (length of the unit lattice along the x and y axes is about two times that in the direction of the z axis, the elastic modulus of single crystal Sn along the crystal axis is about three times of that along the x and y axes, and the coefficient of thermal expansion along the z axis is two times that along the two other axes [10]). The anisotropy of mechanical properties plays an important role in the reliability of solder interconnections because SnAgCu interconnections are typically composed of only a few large grains. Due to the body centric tetragonal structure, Sn is also prone to strain rate hardening.

Finally, it should be pointed out that mechanical properties of all materials, metals in particular, are dependent on their microstructures. To a large extent the mechanical properties of crystalline materials can be attributed to their crystal structure and their imperfections (e.g. point, line, planar defects, voids, second phase particles, etc.) but there are a number of characteristics that should also be considered and grain size is only one of them. The relation commonly known as the Hall–Petch equation shows that the yield strength is inversely proportional to the square root of the average grain size of polycrystalline materials (strength is increased with decreased grain size) [11–14]. This relation points out that the mechanical properties are dependent also on the volume fraction of the interfacial phases and the associated changes in the efficiency of dislocation-based mechanisms to mediate plastic deformation. Therefore over the past few decades, nanocrystalline metals and alloys have been the subject of considerable research. For example, it has been recognised that metals with significantly small grain size (not exceeding the range of 100 nm to 1 μm) have much higher yield and fracture strengths, lower ductility and toughness and superior wear and fatigue resistance, as compared to their microcrystalline counterparts [15, 16]. For the same reasons, the mechanical properties of extremely thin films differ significantly from their macroscopic counterparts [17, 18].

3.2 Restoration of Plastically Deformed Metals

The response of materials to mechanical loading was discussed in the previous section. Deformation is the response of a material to an applied load and the plastic part of deformation involves permanent changes to microstructures. In the case of metals, unlike in the case of most polymers, microstructures are recoverable at least to a certain extent. During restoration the microstructures of plastically deformed materials are gradually repaired and the physical properties such as yield strength, hardness, ductility, thermal conductivity, electrical resistivity and density are gradually restored towards their values before deformation. The driving force of the restoration processes is the release of internal energy bound in the material during deformation.

During deformation a part of the work consumed is stored in the material in the form of lattice defects, primarily in dislocations. The increased internal energy of deformed material acts as the driving force for the two competitive restoration processes: recovery and recrystallization. Recovery and recrystallization are competing processes because they are both driven by the same stored energy. However, the activation energy of recovery is lower than that of recrystallization and, therefore, the onset of recovery is much faster. In fact, recrystallization has an inherent incubation period. Thus, the progress of recovery can decrease the driving force of recrystallization significantly and thereby hinder or can even prevent the onset of recrystallization.

Recovery occurs without notable changes in the microstructure but recrystallization gives rise to clearly visible microstructural changes even within the resolution of the optical microscopy. In recrystallization new strain-free grains are nucleated preferentially at grain or phase boundaries from where they grow in size and gradually consume all the deformed material. The kinetics of recrystallization is dependent, among other things, on the degree of deformation, time and temperature, according to the well-known Avrami equation [19]. The restoration processes are followed by grain growth, where the recrystallized grains grow larger with time owing to the decrease of grain boundary energy.

3.3 Formation of Strains and Stresses in the Electronic Assemblies

Electronic devices are formed by connecting electronic components together with the help of a printed wiring board and solder or adhesive. In their use environments these multimaterial component boards are exposed to various loadings, such as changes of temperature, drops, bending and/or vibration. Under such loading conditions strains and stresses become concentrated at the interfaces between dissimilar materials. Since the increasingly miniaturized solder or adhesive interconnections are located between these two highly dissimilar materials (package and PWB), strains and stresses produced by operation loads are concentrated in these relatively small interconnections, which accommodate stresses by deformation. Therefore electrical failures of electronic component boards are often associated with cracking of the interconnections between electronic components and printed wiring boards because, as pointed out earlier in this chapter, electronic component boards are composed of a variety of different materials that have very dissimilar mechanical properties, such as the coefficient of thermal expansion (CTE) and elastic modulus. Mismatches in mechanical properties of adjoining materials are found everywhere in electronic products and at various length scales. For example, changes in the temperature of electronic products cause strains and stresses (i) on the product level between the component board assembly and product covers or mounting fixtures, (ii) on the assembly level between printed wiring board and components, (iii) at interfaces between different materials and (iv) on the microstructural level between different phases of multicomponent materials. Under frequently repeating conditions these thermomechanically produced strains and stresses can cause assemblies to fail by fatigue. On the other hand, strains and stresses are produced in a similar manner as component boards are bent, twisted or dropped. Under a shock impact caused by an accidental drop of an electronic device, the impact energy is transmitted from the point of impact throughout the device structure by bending and vibration of the structures. As we shall soon point out, there are significant similarities in the formation mechanisms and the distribution of strains and stresses between the component boards under thermomechanical and shock loading conditions but the response of materials, that of solders in particular, to these loads is quite different. The produced stress and strain conditions are highly complex and difficult to comprehend intuitively. Therefore the Finite Element Method (FEM) is often employed in practice to evaluate stresses and deflections in electronic products and component boards.

3.3.1 Electronic Component Boards Assemblies under Changes of Temperature

It is generally recognised that thermomechanical loading is a major cause of failures in electronic products because almost all electronic devices experience changes in temperature, caused by either the external environment or internal heat

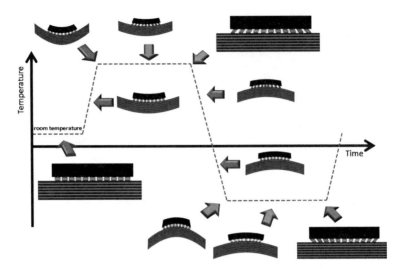

Fig. 3.3 Formation mechanism of strains in electronic assemblies under changes in temperature

production. Thermomechanical stresses are produced when the natural thermal expansion of materials is restricted and the stresses are concentrated at the interfacial regions between adjoining dissimilar materials. In electronic assemblies thermomechanical stresses are produced to different extents at all length scales from submicron intermetallic particles submerged in the Sn matrix of solder interconnections to the structural dimensions of a product. When an electronic component board is undergoing a change in temperature, a mismatch in CTE between adjoining materials causes either deformation of the structures (such as bending or twisting) or increase of internal stresses, depending whether the displacements due to thermal expansion are restricted or not. Figure 3.3 shows schematically how strains and stresses are produced in electronic component assemblies undergoing changes in temperature as the bending of the printed wiring board is relatively unrestricted.

The CTE of most PWBs is much higher than that of most packages. For instance, the CTE of FR4, which is the most commonly used base material of PWBs, is about $16 \times 10^{-6}/°C$ [20] whereas that of silicon is only $2.5 \times 10^{-6}/°C$ [21]. Furthermore, due to the large volume fraction of the silicon, chip scale packages have much higher rigidity that the PWBs and, therefore, the effect of changes in temperature is bending of the PWB and the concentration of strains and stresses are in the solder interconnections between the packages and substrate. When components and board expand by different amounts, the amount of shear strain (γ) can be approximated according to a simple analytical formula: in addition to the magnitude of thermal change (ΔT) and the difference in the CTEs ($\Delta \alpha$), the extent of deformation ($\Delta \gamma$) depends on the structure of the component, namely the height (h) of solder interconnections and the distance of

3.3 Formation of Strains and Stresses in the Electronic Assemblies

Fig. 3.4 Stress–strain hysteresis loop of solder interconnections under isothermal cyclic loading and at thermal cycling [22]. Reprinted by courtesy of IEEE

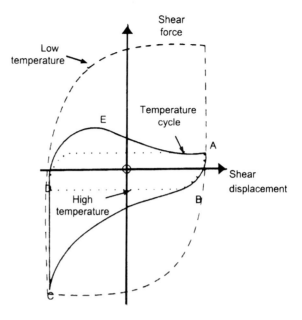

interconnections from the neutral point (L). Their relation can be presented by the equation $\Delta\gamma = \Delta\alpha \times \Delta T \times L/h$. The generated strains and stresses are mainly shear. The division of stress into share and tensile components is important in the reliability analyses of solder interconnections because of the tendency of the shear component to promote deformation and the tensile component to advance crack propagation.

As discussed above, the mechanical properties of materials are dependent on temperature. This fact has particular importance for solder interconnections between packages and PWBs since (a) thermomechanical strains and stresses are concentrated in and accommodated by deformation of the solder interconnections, and (b) the temperature range of operation environment in most of the electronic applications is relatively high (high homologous temperature), which means that creep processes will contribute to plastic deformation of solder interconnections during thermal cycling.

Since the strength and creep rate of solder interconnections is dependent on temperature, the amount of plastic deformation in solder interconnections depends on the temperature region of a thermal cycle. At low homologous temperatures the cyclic stress–strain relationship of solder interconnections constitutes a symmetric hysteresis loop, as shown by the dashed line in Fig. 3.4. The area contained within the hysteresis loop represents the plastic work done on the material, the width of the loop represents plastic strain range and the elastic strain range is given by the difference between the total strain and the plastic strain. At high homologous temperatures the strength of solder interconnections is lower and the creep is much more efficient and the temperatures shape of the cyclic stress–strain relationship changes as shown by the dotted line in Fig. 3.4. However, standardised thermal cycling tests, such as the IEC standard 68-2-14 [23], set the extreme temperatures at −45°C and +125°C, which for the eutectic SnAgCu alloy is equal to 0.5–0.8 in terms of homologous temperature. This means that the difference in the yield strength and creep rate of solder interconnections is notably different at each extreme temperature and as a result the hysteresis loop becomes highly asymmetric with the maximum stress values at high temperatures being appreciably smaller than those at low temperatures, as depicted by the continuous line in Fig. 3.4. Thus, more extensive plastic deformation takes place in the high temperature part of the cycle and deformation is much less pronounced in the lower temperature part. Therefore, at elevated temperatures the bending of the PWB, as depicted in Fig. 3.3, is much less significant as it is at low temperatures.

It has been observed that typically only a fraction of the solder interconnection cross-section actually participates in cyclic deformation because stress distribution inside solder interconnections is seldom uniform. The plastic deformation in the most highly stressed regions of interconnections leads to localised evolution of interconnection microstructures and ultimately leads to cracking through the bulk of the solder interconnections.

There are several approaches to consider when designing electronic devices for an environment where changes of temperature are considerable. The simple analytical equation discussed above ($\Delta\gamma = \Delta\alpha \times \Delta T \times L/h$) points out several aspects. It is obvious that thermomechanical reliability can be improved by decreasing the difference in coefficients of thermal expansion ($\Delta\alpha$) and by applying a method for more efficient cooling (ΔT). Quite often, however, choices related to these alternatives are limited. The ongoing trend of miniaturisation helps as the dimensions of the electronic packages (L) are shrinking. Reliability is also increased with the introduction of more flexible PWB materials and thinner packages with lower stiffness. On the other hand, decreased interconnection densities (decreased h) bring new challenges as the packages become closer to the printed wiring boards. However, apart from these alternatives, the choice of load bearing materials is also something to consider. In the case of solder alloys there are several compositions of Sn-rich lead-free alloys commercially available that have been tailored for better stability of microstructures and fatigue resistance.

3.3.2 Electronic Component Boards under Vibration and Mechanical Shock Loading

Several studies carried out with commercial portable electronic products have shown that impact forces generated when products are dropped onto the ground are transmitted through the product casing to the component boards and make the boards bend and vibrate excessively. The results from product-level tests have

3.3 Formation of Strains and Stresses in the Electronic Assemblies 35

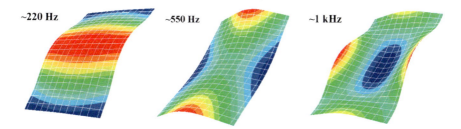

Fig. 3.5 Total flexure of the test board is the sum of the different natural modes. Three of the most significant natural modes of the JESD22-B111 compliant test board. Different shades of grey represent vertical displacement [26]. Reprinted by courtesy of the author

been used to develop board-level drop tests [24, 25]. The condition B of the widely used JESD22-B111 drop test standard defines the deacceleration pulse as having a shape of half-sine with 0.5 ms width and maximum at 1500 Gs. The shape of the pulse is not only a function of the drop height but depends on the characteristics of the strike surface: drop height determines the maximum deacceleration and strike surface the pulse width. The component board is attached to a support fixture from its four corners with the components facing downwards. The fixture is mounted on a sledge that is dropped down to a rigid surface from a specified height in a controlled manner with the help of two guiding rails. Placing the printed wiring board horizontally results in the maximum flexure of the test board.

In the drop tests after the moment of impact the component boards bend downwards forced by the inertia and oscillate at a high frequency for about 20–30 ms. The free vibration of the component board after the drop impact takes place by many different modes simultaneously and the total bending of the component board is their sum. Figure 3.5 illustrates how the component board responds when it is being placed under the JESD22-B111 shock loading conditions.

The shapes of the natural modes depend on the support structure of the component board, whereas the natural (resonant) frequencies depend on the stiffness and mass of the component board. Figure 3.6 shows a strain history measured in the centre of the board layout and in the longitudinal direction of the board. It shows that many vibration modes are initiated simultaneously as each mode produces a component of the total strain history that has the same frequency as the mode and strain levels proportional to the displacement amplitude of the mode. The "macroscopic" oscillation is due to the natural mode with the lowest frequency while oscillations at higher frequencies are summed on the larger strains. The contribution of the natural modes with highest frequencies to the total bending of the board is usually not significant because their amplitude is relatively small and their vibration is attenuated relatively fast. Only the lowest frequencies, in the case of the JESD22-B111 board the lowest three, are considered significant. Owing to the simultaneous action of many different natural modes and frequencies, and the fast attenuation of the vibration amplitude, the strain distribution on

Fig. 3.6 Measured longitudinal strain at the centre of the board

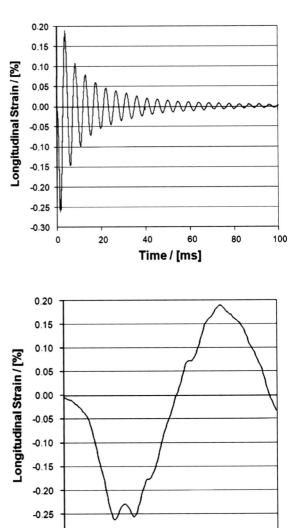

the test board changes very rapidly. Thus, the location of the highest stress changes quickly too but most often components mounted on the centre line parallel to the short edge of the board experience the highest trains and stresses.

Bending causes displacement between the printed wiring board and the components. Stresses concentrate at the interconnection regions where they cause component, solder interconnection, or board failures. If we compare the strains and stresses produced by thermomechanical loads with those produced by the (high-

3.3 Formation of Strains and Stresses in the Electronic Assemblies

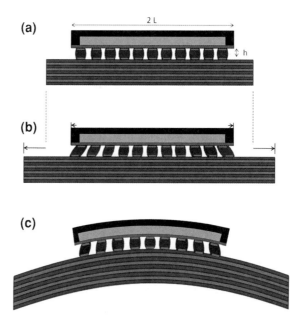

Fig. 3.7 a component board at rest, b displacements caused by thermal cycling, c displacements caused by bending of the PWB (drop testing)

frequency) bending of the printed wiring board (PWB) during the drop tests, there are a number of obvious similarities (see Fig. 3.7). Under thermal cycling, the solder interconnections experience strains and stresses due to the differences in CTE of the component packages and the PWBs. In drop tests, on the other hand, strains and stresses are also concentrated in the solder interconnections but they are caused by the bending of the PWB. However, the significance of the shear components in the thermomechanical stress/strain condition under the thermomechanical loading are dominant while as a consequence of the board bending, the tensile component is dominant in the shock and vibration loading conditions. This minor detail has significant consequences on the failure mechanisms under the two different loading conditions. Readers interested in more thorough discussion on the topic can refer e.g. to [31–33].

The failure mechanism under mechanical shock loading differs greatly from that under thermomechanical loading. Besides temperature the most important difference between drop tests and thermal cycling tests is the deformation rate. As noted earlier in this chapter, at high homologous temperatures (above 0.3–0.4) the deformation of metals becomes time-dependent and the plastic flow depends on the rate of strain. In thermal cycling tests the deformation rate of solder interconnections is in the range of 10^{-6}–10^{-4}/s, while in drop tests maximum strain rates can reach 10–100/s. Both the ultimate tensile strength and the yield strength increase with strain rate, but the yield strength is typically much more sensitive to strain rate [9]. Figure 3.8 shows the ultimate tensile strength of Sn and two common Sn-based solders, Sn1.5Bi and Sn3.4Ag0.8Cu, as a function of strain

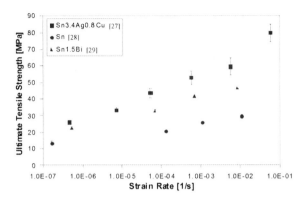

Fig. 3.8 Effect of strain rate on strength of Sn and two Sn-based solders [27–29]

rate. As can be seen, when the strain rate is increased from that occurring in thermal cycling to that occurring in drop tests, the flow stress of solder is about two to three times as high.

The increased strain rates not only increase the stresses in solder interconnections, but they also become more concentrated on the component side of the interconnections. Due to the much higher stress levels in the solder interconnections during the drop tests, the intermetallic compound layers will experience significantly higher stresses than those under thermal cycling conditions. The tensile strength of the solder increases above the fracture strength of the intermetallic layers and this ultimately makes the fractures propagate inside the IMC layers, instead of the bulk solder.

The approaches to design electronic assemblies that are resistant to drop and vibration load are somewhat similar to those for assemblies resistant to changes of temperature. Any decrease in the dimensions of the electronic packages (L) decreases the strains and stresses in the vicinity of solder interconnections while the decrease in interconnection height increases them. However, there is one important distinction related to the choice of the interconnection material. The lower strength of solder interconnection can be beneficial for drop reliability but detrimental to thermal cycling reliability. Furthermore, modification of the intermetallic layers, for example by the introduction of minute amounts of Ni in the SnAgCu solders, changes the morphology of the interfacial reaction layers and, thereby, can make them less susceptible to failure in the drop tests.

3.4 Failures in Electronic Component Boards

The failure mechanisms of many electronic materials, particularly those of solders, are not well defined. Cracking of solder interconnections is particularly complicated because temperatures of common use environments are well above the 0.5 homologous temperatures and, therefore, cracking of solder interconnections is often influenced by marked changes of the as-solidified microstructures.

3.4 Failures in Electronic Component Boards

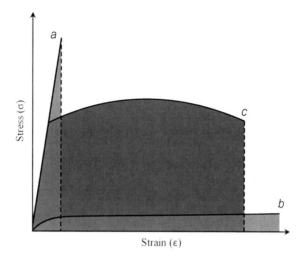

Fig. 3.9 Idealised stress–strain curves of **a** a brittle material (high strength, low ductility), **b** a ductile material (low strength, high ductility), and **c** an elastoplastic material (moderate strength, moderate ductility)

Furthermore, Sn-based solders are also strain rate sensitive and exhibit dissimilar failure modes under different loading rates. The details of solder interconnection failures will be discussed later in this book. Before that, however, we will make a brief review of the common failure mechanisms of electronic materials since these mechanisms may be involved with failures elsewhere in electronic assemblies.

Fracture is often defined as the separation of a solid body into two or more parts under the action of stress and the fracture strength is the stress under which the part fails. There are two distinguishable stages of cracking: nucleation and propagation of cracks. Cracks typically nucleate at dents, pits, voids or other defects on the surface of material. In the case of crystalline materials, on the other hand, cracks propagate transgranularly through their grains or, in the case of polycrystalline materials, they can alternatively propagate intergranularly along grain boundaries.

Failure modes of engineering materials are commonly classified as brittle and ductile based on the amount of energy consumed prior to crack nucleation and during propagation, i.e. the fracture toughness. A ductile fracture is a high-energy fracture because a relatively large amount of energy is absorbed due to the plastic deformation occurring during fracture. A brittle fracture is a low-energy fracture because cracks propagate with little or no plastic deformation. Brittle cracks show very little or no plastic deformation before crack nucleation and the propagation of cracks takes place rapidly. See Fig. 3.9 for graphical presentation. Once nucleated brittle cracks do not need an appreciable increase of applied stress to continue propagation. Ductile cracks demonstrate the opposite. Ductile cracks exhibit large amounts of plastic deformation both before crack initiation as well as during propagation. Specimens that fail by the ductile mode under uniaxial tension exhibit plastic deformation prior to necking throughout the specimen but as the necking begins, plastic deformation becomes concentrated around the neck region and, finally, after a crack is nucleated, the plastic deformation is concentrated at the front of the crack tip. Because of the plastic work involved, the propagation rate of ductile cracks is much slower and the propagation of cracks does not continue unless there is an increase in the applied stress. The tendency towards brittle fracture is increased at low temperatures and high strain rates. The division to brittle and ductile cracking is arbitrary and depends on the situation being considered.

The mechanisms by which materials fail depend, among other things, on temperature, the loading conditions and the loading rate. The failure mechanism tells how cracks nucleate and develop, whereas the failure mode describes the fracture. The basic fracture mechanisms, microvoid coalescence and fatigue fracture, are discussed in the following.

3.4.1 Fracture Modes

Tin has been observed to fail by void coalescence at the grain boundaries at elevated temperatures (190°C) when stressed uniaxially under a constant load [30]. The fracture mechanism of microvoid coalescence is characterised by nucleation of microvoids and their growth and coalescence to form cracks. The nucleation of voids due to plastic straining is typically attributed to either particle cracking, interfacial decohesion between a particle and the surrounding matrix, or decohesion between grains. As the deformation continues, these voids enlarge, which consumes most of the energy required for fracture. The final step of the fracture mechanism is coalescence of the numerous voids in the direction of maximum shear stress and necking down of the ligaments between adjacent microvoids, resulting in the formation of localised cracks. The cracks propagate slowly because they only extend when additional stress is applied. The resultant fracture surface is of highly irregular appearance, and the size of the dimples visible on the fractured surfaces may vary widely because nucleation of microvoids depends on several different factors (including size, stress and strain levels, the amount of deformation and purity of the material, for instance). Microvoid coalescence is a typical failure mechanism of ductile fracture.

3.4.2 Fatigue Failures

It has been recognised that metals can fail under repetitive or fluctuating loading under much lower stress than that required to cause fracture under a single application of load—or even at stress levels below the yield strength. The classical fatigue failures typically occur suddenly and unexpectedly because observable (macroscopic) plastic deformation does not occur before failure. Surface topology and the

3.4 Failures in Electronic Component Boards

shape of the loaded object affect the fatigue life as the presence of scratches, notches, dents or any other similar features can cause localised stress concentrations where fatigue cracks can initiate. Fatigue failure is a three-stage process involving crack nucleation, crack growth and final failure. Fatigue cracks nucleate preferentially at scratches, notches, dents or wherever stresses can concentrate on the surface of a material. When defects or cracks are preexistent, the nucleation stage is shortened or omitted entirely, and the number of cycles to failure is reduced. However, if suitable preexistent sites for crack nucleation are not available, plastic deformation will produce intrusions and extrusions on the surface of the material. This microplasticity occurs preferentially at the surface of a material because the material is constrained everywhere else. Crack nucleates grow slowly with an increasing number of load cycles until the component fails.

Fatigue is commonly divided into two categories: (i) stress controlled fatigue or high cycle fatigue, which is cracking of a material under repeatedly applied stress levels below yield and (ii) strain controlled fatigue or low cycle fatigue, where applied stress exceeds the yield strength of the material. The nucleation of cracks plays an important role in the formation of fatigue cracks. In an ideally "flawless" material there is no nucleation sites readily available prior to loading. If suitable preexistent sites for crack nucleation are not available, plastic deformation will produce intrusions and extrusions on the surface of the material. This microplasticity occurs preferentially at the surface of a material because the material is constrained everywhere else. When an applied load exceeds the yield strength of a material (which is, for example, the case with solder interconnections under thermal cycling loads), the formation of crack nucleates is relatively fast and the number of load cycles to failure remains low (below about 10^4 cycles). However, when the applied load remains below the yield strength of a material, the formation of crack nucleates can take much longer time and the number of load reversals to failure can be significantly high (well above 10^4 cycles). In most environments, an oxide layer very quickly covers the surface of this fresh material at the tip of the crack. The following compressive cycle pushes the crack shut, making the new surface fold forward. This same mechanism repeats during every cycle and the crack propagates during each cycle until the material fails completely. Any inclusions within the plastic zone at the tip of the crack enhance the growth rate. When the cracks grow by a small amount in each cycle, striations characteristic for fatigue fracture are produced on the fractured surfaces. At a certain stage in the growth of a fatigue crack, the area of the uncracked cross-section will be reduced to a point where the stress acting on the remaining surface reaches a level at which an ordinary brittle or ductile fracture can occur.

It should be noted that some materials, such as solder interconnections under thermomechanical loading, fail by fatigue mechanisms but the failure mechanism differs from the classical fatigue failure mechanism because significant plastic deformation and changes of microstructures precede crack nucleation and propagation (see e.g. [34, 35]).

References

1. M.F. Ashby, D.R.H. Jones, *Engineering Materials—An Introduction to Their Properties and Applications* (Pergamon Press, Oxford, 1980), p. 278
2. H. Ma, J. Suhling, A review of mechanical properties of lead-free solders for electronic packaging. J. Mater. Sci. **44**, 1141–1158 (2009)
3. M.E. Fine, in Physical basis for mechanical properties of solders, ed. by K.J. Puttlitz, K.A. Stalter. *Handbook of Lead-Free Solder Technology for Microelectronics Assemblies* (Marcel Dekker, Inc., New York, 2004), pp. 211–237
4. G. Dieter, *Mechanical Metallurgy* (McGraw-Hill, Inc., New York, 1986), p. 751
5. W.D. Callister, *Materials Science and Engineering—An Introduction* (Wiley, New York, 2007), p. 975
6. W.G. Johnston, J.J. Gilman, Dislocation velocities, dislocation densities, and plastic flow in lithium fluoride crystals. J. Appl. Phys. **30**(2), 29–144 (1959)
7. J.J. Gilman, W.G. Johnston, in The origin and growth of glide bands in lithium fluoride crystals, ed. by J.C. Fuisher, W.G. Jognston, R. Thomson, T. Vreeland Jr. *Dislocations and Mechanical Properties of Crystals* (John Wiley & Sons, New York, 1956), pp. 117–163
8. J.P. Hirth, J. Lothe, *Theory of Dislocations* (McGraw-Hill, New York, 1968), p. 780
9. G.E. Dieter, *Mechanical Metallurgy*, 3rd edn. (McGraw-Hill Book Company, New York, 1986), p. 751
10. T. Lyman, H. E. Boyem, P. N. Unterweiser, J. E. Foster, J. P. Hontas, H. Lawton, Properties and Selection of Metals, *Metals Handbook*, vol. 1, 8th edn. (American Society for Metals, New York, 1961), p. 1300
11. W. Sylwestrowicz, E.O. Hall, The deformation and ageing of mild steel. Proc. Phys. Soc., **64**(6), 495–502 (1951)
12. E.O. Hall, The deformation and ageing of mild steel: II characteristics of the Lüders deformation. Proc. Phys. Soc., **64**(9), 742–747 (1951)
13. E.O. Hall, The deformation and ageing of mild steel: III discussion and results. Proc. Phys. Soc., **64**(9), 747–753 (1951)
14. N.J. Petch, The cleavage strength of polycrystals. J. Iron Steel Inst. **5**, 25–28 (1953)
15. K.S. Kumar, H. Van Swygenhoven, S. Suresh, Mechanical behavior of nanocrystalline metals and alloys. Acta Mater. **51**, 5743–5774 (2003)
16. C.P. Pande, K.P. Cooper, Nanomechanics of Hall–Petch relationship in nanocrystalline materials. Prog. Mater. Sci. **54**, 689–706 (2009)
17. S.P. Baker, Plastic deformation and strength of materials in small dimensions. Mater. Sci. Eng. A **16**, 319–321 (2001)
18. W.D. Nix, Mechanical properties of thin films. Metall. Trans. A **20**(11), 2217–2245 (1989)
19. M. Avrami, Kinetics of phase change. II Transformation–time relations for random distribution of nuclei. J. Chem. Phys. **8**, 212–224 (1940)
20. C.F. Coombs Jr. *Printed Circuits Handbook*, 5th edn. (McGraw-Hill, New York, 2001), p. 1200
21. Y.S. Touloukian, C.Y. Ho, *Thermal Expansion: Metallic Elements and Alloys* (IFI/Plenum, New York, 1975), p. 316
22. P.M. Hall, Forces, moments, and displacements during thermal chamber cycling of leadless ceramic chip carriers soldered to printed boards. IEEE Trans. Compon. Hybrids Manuf. Technol. **7**(4), 314–327 (1984)
23. IEC 60068-2-14 Ed. 5.0 b: 1984, Environmental testing—part 2: tests. Test N: change of temperature, International Electrotechnical Commission, (1984), p. 34
24. JESD22-B111, *Board Level Drop Test Method of Components for Handheld Electronic Products*. (JEDEC Solid State Technology Association, Arlington 2003), p. 16
25. IEC 91/530/NP, Surface mounting technology—environmental and endurance test methods for surface mount solder joint. Part 3: cyclic drop test, International Electrotechnical Commission, proposal (26.9.2005), p. 14

References

26. P. Marjamäki, *Vibration Test as a New Testing Method for Studying The Mechanical Reliability of Solder Interconnections under Shock Loading Conditions*, Dissertation, Espoo, 2007, TKK-EPT-17, Otamedia, p. 128
27. T.O. Reinikainen, P. Marjamäki, J.K. Kivilahti, Deformation characteristics and microstructural evolution of SnAgCu solder joints, *The Proceedings of the 6th EuroSimE Conference*, Berlin, Germany, 18–20 April 2005, IEEE, (2005), pp. 91–98
28. R. Nikander, *Characterization of the Mechanical Properties of the Dilute Tin Based Solder Alloys*, Espoo, Master's thesis, Helsinki University of Technology, 1999, p. 79
29. T. Reinikainen, J.K. Kivilahti, Deformation behavior of dilute SnBi(0.5 to 6 at. pct) solid solutions. Metall. Mater. Trans. A **30**, 123–132 (1999)
30. P. Adeva, G. Caruana, O.A. Rauno, M. Torralba, Microstructure and high temperature mechanical properties of tin. Mater. Sci. Eng. A **194**(1), 17–23 (1995)
31. T.T. Mattila, M. Paulasto-Kröckel, Toward comprehensive reliability assessment of electronic component boards. Microelectron. Reliab. (a Special Issue) **51**(6), 1077–1091 (2011)
32. T.T. Mattila, *Reliability of High-Density Lead-Free Solder Interconnections under Thermal Cycling and Mechanical Shock Loading*, Espoo, 2005, HUT-EPT-13, Otamedia, p. 202, http://lib.tkk.fi/Diss/2005/isbn9512279843/
33. T.T. Mattila, T. Laurila, J.K. Kivilahti, in Metallurgical factors behind the reliability of high density lead-free interconnections, ed. by E. Suhir, C.P. Wong, Y.C. Lee. *Micro-and Opto-Electronic Materials and Structures: Physics, Mechanics, Design, Reliability, Packaging*, vol. 1 (Springer Publishing Company, New York, 2007), pp. 313–350
34. T.T. Mattila, J.K. Kivilahti, The role of recrystallization in the failure mechanism of SnAgCu solder interconnections under thermomechanical loading. IEEE Trans. Compon. Packag. Technol. **33**(3), 629–635 (2010)
35. T.T. Mattila, M. Paulasto-Kröckel, J.K. Kivilahti, in The failure mechanism of recrystallization assisted cracking of solder interconnections, ed. by K. Sztwiertnia. *Recrystallization* (Intech Open Access Publishing), ISBN 979-953-307-346-9, (in print)

Chapter 4
Introduction to Thermodynamic-Kinetic Method

In order to have a better understanding of the basic failure modes and the underlying mechanisms of microsystems under different stress states, more attention should be paid to the chemical compatibility between dissimilar materials found at various levels of devices [1]. When two materials are in contact interdiffusion, chemical reactions etc. can take place across the interface. The energetic prerequisites of different phenomena, which can occur at the interface, can be evaluated by thermodynamics. When this information is supplemented by diffusion kinetic considerations, evolution of the interfacial reaction zone as a function of time can be evaluated. This is the very nature of the thermodynamic-kinetic (T-K) method. In this chapter the basics of the T-K-method will be presented.

4.1 Thermodynamics

The thermodynamics of materials provides fundamental information both about the stabilities of phases and on the driving forces for chemical reactions and diffusion processes, even though complete phase equilibrium (global) is hardly ever met in electronics applications. However, stable or metastable local equilibria are generally attained at the interfaces. Thus, the binary, ternary and higher order phase diagrams, as well as other stability diagrams, can be employed for determining the possible reaction products and this provides a feasible method—together with kinetic information—for analyzing the interfacial reactions in interconnection systems. Next, some definitions which are needed during the subsequent treatment of the subject are introduced. *Component* refers to independent species in the system under investigation. Thus, it gives the minimum number of substances, which must be available in the laboratory in order to make up any chosen equilibrium mixture of the system in question. The *system* is a clearly defined part of a macroscopic space, distinguished from the rest of the space by a clear boundary. The rest of the space (taking only the part that can be regarded to interact with the system) is defined as the *environment*. The system can be *isolated, closed or open* depending on its interactions with the environment. An isolated system cannot exchange energy or matter, a closed system can exchange energy, but not matter and an open system can exchange both energy and matter with the environment. A *phase* is a region of uniformity in a system under investigation. It is a region of uniform chemical composition and uniform physical properties. A phase is also distinguished from other dissimilar regions by an interface. In thermodynamics one has *extensive, intensive* and *partial* properties. Extensive properties depend on the size of the system whereas the intensive properties do not. Partial properties are the molar properties of the component. Formal definitions of the laws of thermodynamics are not considered here and the reader is referred to the literature for a more extensive treatment of the topic [2–8]. In the subsequent chapters, the ingredients needed to realise the thermodynamic-kinetic method are introduced. In addition, some issues connected with the microstructural aspects related to the practical use of the method are discussed. It is to be noted that the concept of microstructure is not always unambiguously defined. Typically, microstructure is taken to be the structure observed under the microscopes. The size scales of structural features of metals, for example, extend from the atomic level (~ 0.1 nm) to the size of the entire metallic object (~ 1 m). Thus, it is evident that different features of the microstructure depend on the observation method used. In this chapter we define the microstructure as the one observed when all possible analytical tools, such as optical microscopy (OM), scanning electron microscopy (SEM), transmission electron microscopy (TEM), X-ray diffraction (XRD) etc. have been utilized to obtain the required information.

4.1.1 Gibbs Free Energy and Thermodynamic Equilibrium

The combined statement of the first and second laws of thermodynamics can be stated in terms of the Gibbs free energy function as follows:

$$G \equiv H - TS, \qquad (4.1)$$

where *H*, and *S* are the enthalpy and the entropy of the system, respectively. Given that $G = G(T, P, n_1, n_2, \ldots)$ in an open system, with n_i being the number of moles of component *i*, the derivative of the Gibbs energy function yields:

$$dG = -SdT + Vdp + \sum_i \mu_i dn_i \qquad (4.2)$$

where μ_i is the chemical potential of component *i*. At a constant value of the independent variables *P*, *T* and n_j ($j \neq i$) the chemical potential equals the partial molar Gibbs free energy, $(\partial G/\partial n_i)_{P,T,j \neq i}$. The chemical potential (partial Gibbs energy) has an important function analogous to temperature and pressure. A temperature difference determines the tendency of heat to flow from one body into

4.1 Thermodynamics

another, and a pressure difference on the other hand, the tendency towards a bodily movement. A chemical potential can be regarded as the cause of a chemical reaction or the tendency of a substance to *diffuse* from one phase to another. Chemical potential can be defined formally as follows [2]: "If to any homogeneous mass we suppose an infinitesimal quantity of any substance to be added and its entropy and volume remaining unchanged, the increase of the energy of the mass divided by the quantity of the substance added is the (chemical) *potential* for that substance in the mass considered".

With the help of the Gibbs free energy function the equilibrium state of the system can be investigated. It is to be noted that because the Gibbs energy is a property of the system only, it provides far more convenient criteria for direction of change than the otherwise more general entropy criterion. In the latter approach, changes in the system *as well as in the environment* must always be considered and the environment is not always easily defined. In addition, as there is the relation between chemical potential of components and the total Gibbs energy of the system through:

$$G_{tot} = \sum_{\phi} \sum_{i} \mu_i^{\phi} n_i^{\phi} \tag{4.3}$$

Gibbs energy function can be utilised from the component level to the system level and back again. Hence Eq. 4.3 provides the very important bridge between component and system level properties.

Three stable equilibrium states to be considered are (i) complete thermodynamic equilibrium, (ii) local thermodynamic equilibrium and (iii) partial thermodynamic equilibrium. When the system is at complete equilibrium its Gibbs free energy (G) function has reached its minimum value:

$$dG = 0 \text{ or } \mu_i^{\alpha} = \mu_i^{\beta} = \ldots = \mu_i^{\phi}, \ (i = A, B, C, \ldots) \tag{4.4}$$

and then the system is in mechanical, thermal and chemical equilibrium with its surroundings. Thus, there are no gradients inside the individual phases. Owing to this, no changes in the macroscopic properties of the system are expected.

On the other hand, local equilibrium is defined in such a way that the equilibrium exists only at the interfaces between different phases present in the system. This implies that the thermodynamic functions are continuous across the interface and the compositions right at the interface are very close to those indicated by the equilibrium phase diagram. It also means that there are activity gradients in the adjoining phases. These gradients, together with diffusivities, determine the diffusion of components in various phases of a joint region. Since the complete thermodynamic equilibrium is seldom achieved in microsystems, the concept of the local equilibrium is central for the analysis of interfacial reactions.

Partial equilibrium means that the system is in equilibrium only with respect to certain components. It is generally found that some processes taking place in the system can be rapid while others are relatively slow. If the rapid ones occur fast enough to fulfil the requirements for stable equilibrium (within the limit of error) and the slow ones slow enough that they can be ignored, then it is quite proper to treat the system as being in equilibrium with respect to the rapid processes alone [3].

It is also possible that the global energy minimum of the system is not accessible owing to different restrictions. In such a case one is dealing with metastable equilibrium, which can be defined as a local minimum of the total Gibbs energy of the system. In order to obtain the global stable equilibrium activation (e.g. thermal energy) must be brought into the system. It is to be noted that metastable equilibrium can also be complete, local or partial; out of these the local metastable equilibrium concept will be used frequently in the following chapters. Very often one or more interfacial compounds, which should be thermodynamically stable at a particular temperature, are not observed between two materials and then these interfaces are in local metastable equilibrium. Another situation commonly encountered occurs in solid/liquid reaction couples, where during the few first seconds, the solid material is in local metastable equilibrium with the liquid containing dissolved atoms, before the intermetallic compound(s) is formed at the interface. In fact, a principle commonly known as *Ostwald's rule* states that, when a system undergoing reaction proceeds from a less stable state, the most stable state is not formed directly but rather the next more stable state is formed, and so on, step by step until (if ever) the most stable is formed. It is a fact that most materials used in everyday life have not been able to reach their absolute minimum energy state and are therefore in metastable equilibrium. It should be noted that a system at metastable equilibrium has thermodynamic properties, which are exactly determined, just as a system at stable equilibrium.

The types of metastability in alloys have been classified by Turnbull [9] as compositional, structural and morphological. The degree of metastability is characterised by the excess free energy of the system over that of the equilibrium state. Turnbull [9] expressed this energy per mole as a fraction of $R\bar{T}_m$, where R is the gas constant and $\bar{T}_m$ is the average of the melting points of the elements constituting the alloy. Compositional metastability is where an equilibrium phase exists outside its normal composition range, and it is associated with large excess energies of up to about 1.0 $R\bar{T}_m$. Structural metastability is where a phase has a non-equilibrium structure, typically with excess energies of about 0.5 $R\bar{T}_m$. Morphological metastability has the lowest excess energy, about 0.1 $R\bar{T}_m$, but is perhaps the most widespread, and generally useful, type of metastability. Its excess energy arises from the abnormally large area of grain boundaries and interphase interfaces.

A system in true metastable equilibrium would not have access to any state of lower free energy by means of a continuous structural change [10]. A good example of such a state is a fully relaxed amorphous phase, in which any transformation to a lower free energy microstructure can commence only with a discrete nucleation stage which has an energy barrier. On the other hand, most morphologically metastable microstructures are not in metastable *equilibrium* but can evolve continuously on annealing. In this case, the energy barrier is not nucleation, but the activation energy of atoms making diffusional jumps. Such a system is thermodynamically unstable but is configurationally frozen [10]. Indeed virtually all so-called metastable microstructures, whether in fact metastable or unstable, have technologically useful lifetimes because they are at sufficiently low temperatures to be configurationally frozen.

Finally, it must be emphasised that when the Gibbs free energy function corresponding to the equilibrium state (either stable or metastable) of the system is defined, all other equilibrium properties of the system are also fixed. This enables the extrapolation of large amounts of thermodynamic data, such as activities of the components, chemical potentials of the components, heat capacities, enthalpy, entropy and so forth, from the Gibbs free energy function. These data in turn provide a lot of information that can be used to describe the system thermodynamically in the form of phase diagrams, stability diagrams, predominance diagrams *etc.*

4.1.2 The Chemical Potential and Activity in a Binary Solid Solution

It is common knowledge that most materials in nature consist of several phases, and that the phases are themselves never pure elements. In fact, based on the second law of thermodynamics, a pure substance exists only in our minds and represents a limiting state, which we may asymptotically approach but never actually obtain. Thus, the thermodynamic description of multicomponent systems is of great importance from the theoretical as well as from the practical point of view. In the treatment of multicomponent open systems, the most common process considered in defining the thermodynamic functions for a solution is called the mixing process, which is defined according to Guggenheim [4] as follows: "The mixing process is the change in state experienced by the system when appropriate amounts of the "pure" components in their reference states are mixed together forming a homogeneous solution brought to the same temperature and pressure as the initial state". It is to be noted that although the mixing process is strongly influenced by interaction forces between atoms and molecules (i.e. ΔH_m), the fundamental cause behind mixing is the entropy (ΔS_m) change of the system.

For any heterogeneous system at equilibrium, the chemical potential of a component i has the same value in all phases of the system, where the component has accessibility. An example of a binary phase diagram with limited mutual solubility of the components is presented in Fig. 4.1a. Between the homogeneous terminal phases α and β there is a two-phase region, where α and β coexist. The corresponding Gibbs free energy diagram is shown in Fig. 4.1b. It should be noted that both phases α and β have their individual molar Gibbs free energy curves.

A general problem for dealing with solutions thermodynamically can be regarded as one of properly determining the chemical potentials of the components. Usually the treatment utilises the activity function introduced by Lewis and Randall [5]. Its value lies in close relation to composition; with appropriate choice of reference state the activity approaches the mole fraction as the mole fraction

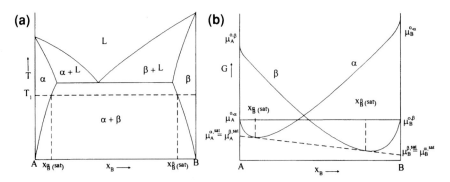

Fig. 4.1 a Binary phase diagram with limited mutual solubility of components A and B, **b** the Gibbs energy diagram at temperature T_1

approaches unity. Most commonly in the thermodynamics of solutions it is not the activity, which is used, but rather the activity coefficient, which is defined as the ratio of the activity a_i to the mole fraction x_i:

$$\gamma_i = \frac{a_i}{x_i} \tag{4.5}$$

In terms of the chemical potential the activity can be expressed as:

$$\mu_i^j - \mu_i^o = RT \ln a_i^j + RT \ln \gamma_i^j \tag{4.6}$$

where μ_i^o is the chemical potential of pure i in the reference or standard state, μ_i^j the chemical potential of i in phase j, a_i^j the activity of component i in phase j, R is the gas constant, T the temperature, and ($i = A,B,\ldots; j = \alpha,\beta,\ldots$). In the limiting case of ideal solutions, where the enthalpy ($\Delta H_m = 0$) and volume change ($\Delta V_m = 0$) of mixing are zero and the only contribution to Gibbs free energy of mixing arises from the configurational entropy term:

$$\Delta S_m = \sum_{i=A}^{K} x_i \ln x_i \tag{4.7}$$

the activity coefficient in Eq. 4.5 is unity and the activity of the component equals its mole fraction (i.e. Raoultian behaviour, see discussion below). If the equality is valid for all compositions the solution is called perfect. Thus, the activity coefficient represents deviation of the real solutions from this limiting behaviour. The use of activity coefficient instead of activity in Eq. 4.6 clearly indicates the excess energy term $RT \ln \gamma_i^j$ responsible for the non-ideal behaviour. This issue is addressed in more detail in Sect. 4.1.5.

The values of chemical potentials μ_A^j and μ_B^j are graphically found at the endpoints of the tangents of the curves α and β in the Gibbs free energy plot (Fig. 4.1b). Inside the two-phase field the values of the chemical potentials are

4.1 Thermodynamics

constant (at equilibrium) and can be found from the tangential points of the common tangent drawn in Fig. 4.1b. In a ternary system, the phase equilibria are obtained by the common tangent-plane construction, where the intersections with the three corners of the diagram represent the chemical potentials. Further details can be found for example from [6].

As only relative values of thermodynamic functions can be determined, an agreed reference state has to be established for each element or species in order to make thermodynamic treatment quantitative. In principle, the choice of the reference state is arbitrary as long as the chosen state is used consequently throughout the analysis. The chosen state is then defined to be zero (in Fig. 4.1b $\mu_A^{o,\alpha}$ and $\mu_A^{o,\beta}$) and all other possible states of the element are compared against the reference state to obtain their relative stabilities. It should be noted that there are some uncertainties related to the usage of reference states in the literature. This issue will be further discussed in Sect. 4.1.5 when thermodynamic calculations are presented.

4.1.3 Driving Force for Chemical Reaction

A chemical reaction taking place at a constant temperature and pressure can be represented by:

$$aA + bB + \cdots \leftrightarrow cC + dD + \cdots \tag{4.8}$$

where A, B, C, etc. are species (reactants and products) in the reaction and a, b, c, etc. are the number of moles of the species in question. The Gibbs free energy change $(\Delta_r G)$ of the reaction is given by the difference in the chemical potentials of the reactants and the products and is defined as [7]:

$$\Delta_r G = (c\mu_C + d\mu_D + \cdots) - (a\mu_A + b\mu_B + \cdots) \tag{4.9}$$

If the change in the Gibbs free energy of the reaction is negative, the reaction can proceed spontaneously unless there are kinetic barriers that hinder the reaction. By giving the Gibbs free energies with respect to the standard states of the species (i.e. $G = G^\circ + RT \ln a$), Eq. 4.8 can be rewritten as:

$$\Delta_r G - \Delta_r G^o = (c\mu_C + d\mu_D + \cdots) - (a\mu_A + b\mu_B + \cdots) \tag{4.10}$$

where $\Delta_r G^o$ is the change in the standard Gibbs free energy of the reaction and $\mu_{A,B,C,D}$ are the chemical potentials of the species A, B, C, D.... Furthermore, Eq. 4.10 can be rewritten with the help of Eq. 4.6 as:

$$\Delta_r G - \Delta_r G^o = RT \ln \left[\frac{a_C^c a_D^d \ldots}{a_A^a a_B^b \ldots} \right] \tag{4.11}$$

At equilibrium $\Delta_r G = 0$. Equation 4.11 is also valid under non-equilibrium conditions. For example, it is applicable when chemical reactions occur in supersaturated metastable liquid solder. In the Sn–Cu–Ni system the chemical reactions to form the IMCs that can be in local equilibrium with the liquid Sn at soldering temperatures are:

$$(6-y)\text{Cu} + y\text{Ni} + 5\text{Sn} \Rightarrow (\text{Cu,Ni})_6\text{Sn}_5$$
$$(3-y)\text{Ni} + y\text{Cu} + 4\text{Sn} \Rightarrow (\text{Ni,Cu})_3\text{Sn}_4 \qquad (4.12)$$

where y is the fraction of Ni atoms in the Cu sublattice in Cu_6Sn_5 and the fraction of Cu atoms in the Ni sublattice in Ni_3Sn_4, respectively. In order to calculate the effect of the activities of dissolved Cu and Ni on the stabilities of the intermetallic compounds, the following equations for the Gibbs energy change of the chemical reactions can be derived:

$$\Delta G^{(\text{Cu,Ni})_6\text{Sn}_5} = (1-y) \times \Delta^0 G^{\text{Cu}_6\text{Sn}_5} + y \times \Delta^0 G^{\text{Ni}_6\text{Sn}_5} + RT \ln\left(\frac{a(\text{Cu, Ni})_6\text{Sn}_5}{a_{\text{Sn}}^5 \times a_{\text{Cu}}^{6-y} \times a_{\text{Ni}}^y}\right)$$

$$\Delta G^{(\text{Ni,Cu})_3\text{Sn}_4} = (1-z) \times \Delta^0 G^{\text{Ni}_3\text{Sn}_4} + z \times \Delta^0 G^{\text{Cu}_3\text{Sn}_4} + RT \ln\left(\frac{a(\text{Ni, Cu})_6\text{Sn}_4}{a_{\text{Sn}}^4 \times a_{\text{Ni}}^{3-z} \times a_{\text{Cu}}^z}\right)$$

$$(4.13)$$

From these equations it can be seen that if the activity of one reactant is very small in the liquid solder and if the activity of IMC is assumed to be 1, the value in brackets becomes large and thus the driving force for the reaction becomes smaller (or even vanishes). The treatment can be further utilised when alloying and impurity effects on interfacial reaction are considered.

4.1.4 The Phase Rule and Phase Diagrams

The phase rule defines the condition of equilibrium in a heterogeneous system by the relation between the number of co-existing phases p, the number of components c and the number of degrees of freedom f as follows:

$$p + f = c + 2 \qquad (4.14)$$

The number 2 in Eq. 4.14 represents pressure and temperature. The number of degrees of freedom of the equilibrium state is the number of conditions (variables) that must be fixed in order to define the corresponding state. It is assumed that temperature, pressure and composition are the only variables that can influence the phase equilibria. For example the effect of surface-tension forces at the boundaries between phases, of gravitational fields, magnetic fields, stresses etc. is considered to be of negligible importance. In many cases the pressure is assumed to be fixed and the phase rule reduces to:

$$p + f = c + 1 \qquad (4.15)$$

4.1 Thermodynamics

A generalised phase rule has been derived for system including surfaces and interfaces by Defay and Prigogine [8]. If c is the number of components, p is the number of 3D phases, φ the number of 2D surface phases, and the number of degrees of freedom f can be expressed as follows:

$$f + p = (c + 2 - p) - (\varphi - s) = v - (\varphi - s) \tag{4.16}$$

where v is the classical (Gibbs) degree of freedom (ignoring surfaces) and s is the number of "surface species". Two surfaces are of different species if they separate different couples of bulk phases. For instance, in the case of a grain boundary precipitate β in a matrix α, the interfaces α/α and α/β are of different species and $s = 2$. Also surfaces of different orientations should be considered as different species. The phase rule provides an important tool when, for example, the correctness of a given phase diagram is validated.

4.1.5 Thermodynamics of Phase Diagrams

In the following, thermodynamic background to understand phase diagrams is briefly introduced. After that binary and ternary phase diagrams are discussed. More complete treatment can be found, for example from [11–13].

4.1.5.1 Free Energy of Solutions

In Sect. 4.1.2 we briefly discussed the so-called mixing process as well as the binary solution phases. A statistical approach can be used to provide more insight into the properties of these phases. The simplest model is one in which the total energy of the solution is given by a summation of interactions between nearest neighbour atoms. If we have a binary system with two types of atoms (A and B) there will be three interaction energy terms. These are the energy of A–A pairs, that of B–B pairs and that of A–B pairs. Here, we assume that the total energy of the solution arises from the interactions between the nearest neighbours. The binding energy may be defined by considering the change in energy as the distance between a pair of atoms is decreased from infinity to an equilibrium separation. The change in energy during this process is the binding energy, which for a pair of A atoms is given as $-2\varepsilon_{AA}$, for B atoms as $-2\varepsilon_{BB}$ and so forth.

The simplest model for real solution phases based on the above defined nearest neighbour interaction approach is the so-called regular solution model. It is based on the following assumptions: (i) mixing among accessible lattice spaces are completely random (ii) atoms interact only with their nearest neighbors, (iii) there is no change in volume upon mixing and (iv) heat capacity does not depend on composition. The energy associated with the mixing process (per interchange) can be described as:

$$dE = 2z\left[\varepsilon_{AB} - \frac{1}{2}(\varepsilon_{AA} + \varepsilon_{BB})\right] = 2ZI_{AB} \quad (4.17)$$

where I_{AB} (per bond) is the interchange energy.

As the energy change associated with the formation of mixture with N_{AB} bonds is $\Delta E = N_{AB}I_{AB}$, the internal energy of the solution phase is:

$$E = \left(\frac{z}{2}\right)N_A(^o\varepsilon_{AA}) + \left(\frac{z}{2}\right)N_B(^o\varepsilon_{BB}) + N_{AB}I_{AB} \quad (4.18)$$

The next step is to identify what the most probable number of AB-bonds (N_{AB}) is with the nominal composition of x_B^o. This problem can be resolved by utilising the first assumption of the regular solution model, i.e. that the mixing among the lattice sites is completely random. The probability then turns out as:

$$\bar{p}_{AB} = z\left(\frac{N_A N_B}{N_A + N_B}\right) \quad (4.19)$$

and therefore the internal energy of the solution phase is:

$$E = \left(\frac{z}{2}\right)N_A(^o\varepsilon_{AA}) + \left(\frac{z}{2}\right)N_B(^o\varepsilon_{BB}) + z\left(\frac{N_A N_B}{N_A + N_B}\right)I_{AB} \quad (4.20)$$

By assuming that $^o\mu_A \cong \frac{1}{2}z(^o\varepsilon_{AA})$ and $^o\mu_B \cong \frac{1}{2}z(^o\varepsilon_{BB})$ and utilising Eq. 4.7 the Gibbs energy of the regular solution phase becomes:

$$G = N_A^o\mu_A + N_B^o\mu_B + kT\sum_i N_i \ln\left(\frac{N_i}{N_A + N_B}\right) + z\left(\frac{N_A N_B}{N_A + N_B}\right)I_{AB} \quad (4.21)$$

from which the molar Gibbs energy is obtained as:

$$g = x_A^o\mu_A + x_B^o\mu_B + RT\sum_i \ln x_i + L_{AB}x_A x_B \quad (4.22)$$

where L_{AB} ($= zNI_{AB}$) is the molar interaction energy, i.e. the *interaction parameter*. By utilising Eqs. 4.5, 4.6 and 4.22 and the fact that $g = \Sigma x_i\mu_i$ we obtain the relationship between activity and the interaction parameter as:

$$a_i = \gamma_i x_i = x_i \exp\left[\frac{L_{ij}(1 - x_i)^2}{RT}\right] \quad (4.23)$$

Hence, the sign of the interaction parameter determines whether the formation of mixture is favoured or hindered. When $\varepsilon_{AA} + \varepsilon_{BB} < 2\varepsilon_{AB}$, and the interaction parameter is negative, the solution will have a larger than random probability of bonds between unlike atoms and thus mixing or compound formation is favoured. The converse is true when the interaction parameter is positive ($\varepsilon_{AA} + \varepsilon_{BB} > 2\varepsilon_{AB}$) since atoms then prefer to be neighbours to their own kind and form clusters. From Eq. 4.23 it is also seen how the activity coefficient depends on both the sign and

4.1 Thermodynamics

magnitude of the interaction parameter. Activity is eventually determined by the interactions between different types of atoms in the solution phase.

In Fig. 4.2 the effect of the sign and magnitude of the interaction parameter (here marked as α) on the formation of a solution phase is shown [12]. When there is no preferred interaction between the atoms in the system the interaction parameter $\alpha = 0$ ($\varepsilon_{AA} + \varepsilon_{BB} = 2\varepsilon_{AB}$) and the integral heat of mixing is zero and the free energy of mixing is given by curve I. As the interaction parameter is made more positive, it can be seen how the enthalpy of mixing becomes more positive and the free energy of mixing becomes less negative. When a certain magnitude in positive interaction is reached (here $\alpha = 2$) it can be seen that the system is about to enter the state where the solution phase becomes unstable. When α is increased to even more positive values one can see how the Gibbs energy curve changes its sign of curvature at the middle region and the so-called miscibility gap is formed. This is associated with the formation of two separate phase regions—one rich in A and another rich in B.

It was discussed above that one may find a situation where the interaction parameter is zero and there is no net interaction between A and B (i.e. $\varepsilon_{AA} + \varepsilon_{BB} = 2\varepsilon_{AB}$). This type of behaviour is associated with the so-called ideal behaviour and is described by Raoult's law. The Raoultian behaviour was defined above as one where the activity coefficient (Eq. 4.5) is unity and the activity of the component equals its mole fraction. Such a behaviour is shown in Fig. 4.3. It was already stated that if Raoult's law is obeyed by the solution phase through the whole composition range the solution is called perfect. This type of solution does not exist in reality, but it provides a convenient reference state to which the behaviour of real solutions can be compared. In Fig. 4.3 another limiting law (Henry's law) is also shown. This limiting law can be understood by utilising the regular solution model and equation [23]. When one approaches the limit where $x_i \to 0$ i.e. the solution becomes dilute, it can be seen from Eq. 4.23 that the activity coefficient becomes concentration independent as:

$$^{\infty}\gamma_i = \exp\left[\frac{L_{ij}}{RT}\right] \quad (4.24)$$

This defines the Henry's law line seen in Fig. 4.3. The limiting laws shown in Fig. 4.3 provide the reference states to which one can compare real solutions. As we approach pure substance ($x_i \to 1$) the solution behaviour necessarily approaches Raoultian behaviour no matter how "non-ideally" it otherwise behaves. This is true also for Henry's law, as all solutions approach it as the solution becomes dilute enough. It is also to be noted that if the solute follows Henry's law then the solvent necessarily follows Raoult's law. Furthermore, whereas perfect solutions do not exist, ideal solution behaviour is commonly encountered in practice within restricted composition limits.

Historically, activity measurements have been carried out mainly by measuring the changes in the partial pressure of a given substance upon alloying with respect to the values of the pure component. As this type of approach also gives an easily

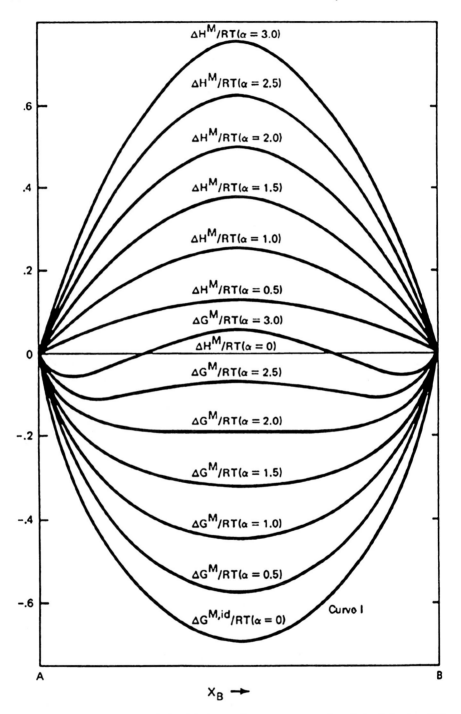

Fig. 4.2 The effect of the magnitude of the interaction parameter α on the solution behaviour [12]

4.1 Thermodynamics

Fig. 4.3 Graphical presentation of Raoult's and Henry's laws

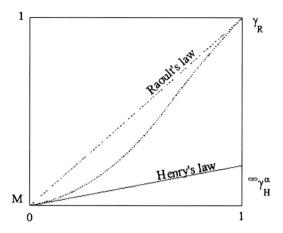

accessible alternative route to derive the above defined limiting laws we shall briefly consider Raoult's and Henry's laws from this point of view. Consider a pure liquid A in a closed vessel (initially evacuated) at temperature T. It will spontaneously evaporate until the pressure in the vessel is equal to the saturated vapour pressure of liquid A (p_A^o) at temperature T. At this point, the rate of evaporation $r_{e(A)}$ and the rate of condensation $r_{c(A)}$ are equal. In order for an atom to escape the surface of the liquid and enter the gas phase it must overcome the attractive forces exerted on it by its neighbours (i.e. overcome the activation energy E* barrier). The magnitude of E* determines the intrinsic evaporation rate. The condensation rate is proportional to the number of A atoms in the vapor phase, which strike (and stick) the liquid surface in unit time. For a fixed temperature the condensation rate is proportional to the pressure of the vapour $r_{c(A)} = kp_A^o$ which is equal to $r_{e(A)}$ at equilibrium. A similar situation holds fo a liquid B. If we now add a small amount of liquid B to liquid A, what happens? If the mole fraction of A in the resulting binary mixture is X_A and assuming that the atomic diameters of A and B are comparable and there is no surface excess, the fraction of the surface area occupied by A atoms is X_A. It is a natural assumption that atom A can evaporate only from a site where it is present and, therefore, $r_{e(A)}$ is decreased by a factor of X_A and the equilibrium pressure exerted by A is decreased from p_A^o to p_A:

$$r_{e(A)} X_A = kp_A \tag{4.25}$$

and by utilising the above defined equality between evaporation rate and equilibrium pressure one obtains:

$$p_A = X_A p_A^o \tag{4.26}$$

which is Raoult's law. A similar equation holds for component B. The law states that the vapor pressure exerted by a component i in a solution is equal to the product of the mole fraction of i in solution and the vapor pressure of i at the temperature of the solution.

While deriving Raoult's law it was assumed that there is no change in the intrinsic evaporation rates. This requires that magnitudes of A–A, B–B and A–B interactions are balanced so that the depth of the potential energy well of an atom at the surface site is independent of the types of atoms surrounding it (see discussion below). If we take that the A–B interaction is much stronger than that between identical atoms and consider a solution of A in B which is sufficiently dilute in such a way that every A atom is surrounded only by B atoms. In this case the activation energy for an A atom to evaporate from the surface is higher than without B and thus the intrinsic evaporation rate will be smaller $\left(r'_{e(A)} < r_{e(A)}\right)$ and equilibrium occurs when:

$$r'_{e(a)} X_A = k p_A \tag{4.27}$$

This results in:

$$p_A = \frac{r'_{e(A)}}{r_{e(A)}} X_A p_A^o \tag{4.28}$$

and as $\left(r'_{e(A)} < r_{e(A)}\right)$ p_A is a smaller quantity than that in Eq. 4.26. Equation 4.28 can be written as:

$$p_A = k'_A X_A \tag{4.29}$$

If X_A of the solution is increased it becomes more probable that not all of the A atoms at the surface are surrounded only by the B atoms. This will have an effect on the activation energy (depth of the potential energy well) and thus after a certain critical value of X_A the intrinsic evaporation rate becomes composition dependent and Eq. 4.29 no longer holds. The above Eq. 4.29 is, of course, Henry's law (similar equation holds for B atoms also). Note also that Raoultian and Henrian activity coefficients have different reference states. More information about the use of these standard states as well as changing between them can be found, for example, from refs. [3, 12].

4.1.5.2 Free Energy of Phase Mixtures

When two materials, especially metals, are mixed together they either form a homogeneous solution or separate into a mixture of phases. Let us consider an alloy X in Fig. 4.4 in the binary system A–B to separate into a mixture of two phases α and β (under a particular temperature and pressure). We shall assume that there are N atoms of alloy X and the fraction of atoms in the α phase is $(1-x)$ and in the β phase x. The number of B atoms in alloy X is n_B^X, the number of B atoms in α-phase is n_B^α and in the β-phase n_B^β. We can change these to atomic fractions by dividing by the total number of atoms N to get:

4.1 Thermodynamics

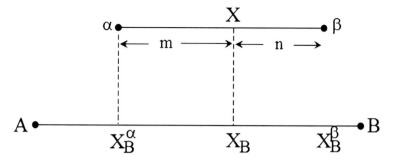

Fig. 4.4 Separation of alloy X into α and β phases

$$X_B = \frac{n_B^x}{N}$$
$$X_B^\alpha = \frac{n_B^\alpha}{N(1-x)} \qquad (4.30)$$
$$X_B^\beta = \frac{n_B^\beta}{Nx}$$

Since $n_B^x = n_B^\alpha + n_B^\beta$
Then

$$NX_B = NX_B^\alpha(1-x) + NX_B^\beta x \qquad (4.31)$$

Whence

$$x = \frac{X_B - X_B^\alpha}{X_B - X_B^\beta} = \frac{m}{m+n} \qquad (4.32)$$

And

$$1 - x = \frac{X_B^\beta - X_B}{X_B^\beta - X_B^\alpha} = \frac{m}{m+n} \qquad (4.33)$$

And

$$\frac{x}{1-x} = \frac{m}{n} \qquad (4.34)$$

This last Eq. 4.34 is called the lever rule, which enables one to calculate relative amounts of phases in a phase mixture in terms of composition of the alloy and phases into which it separates. The free energy of a phase mixture can also be determined by using the lever rule. If alloy X separates into phases α and β, the free energy of an alloy will be unchanged by the separation. The free energy of alloy X is therefore equal to the sum of the free energies of the α and β phases. Since

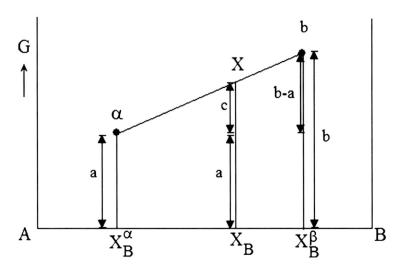

Fig. 4.5 Free energy of a phase mixture

alloy X consists of an amount of α phase equal to $\frac{(X_B^\beta - X_B)}{(X_B^\beta - X_B^\alpha)}$ and similarly $\frac{(X_B - X_B^\alpha)}{(X_B^\beta - X_B^\alpha)}$ of phase β, the free energy of alloy X will be:

$$a\left(\frac{X_B^\beta - X_B}{X_B^\beta - X_B^\alpha}\right) + b\left(\frac{X_B - X_B^\alpha}{X_B^\beta - X_B^\alpha}\right) \quad (4.35)$$

where a and b represent the free energies of the α and β phases at the given temperature and pressure (Fig. 4.5). We can further rearrange Eq. 4.35 to obtain for the free energy of the alloy X:

$$= a\left(\frac{\left(X_B^\beta - X_B^\alpha\right) - \left(X_B - X_B^\alpha\right)}{X_B^\beta - X_B^\alpha}\right) + b\left(\frac{X_B - X_B^\alpha}{X_B^\beta - X_B^\alpha}\right)$$

$$= a + (b - a)\left(\frac{X_B - X_B^\alpha}{X_B^\beta - X_B^\alpha}\right) \quad (4.36)$$

$$= a + c$$

Hence, the alloy X which separates into two phases of composition X_B^α and X_B^β with free energies a and b has a free energy given by the point x on the straight line connecting α and β.

As an example of the use of common tangent construction, the calculation of two phase equilibrium is presented (Fig. 4.6). The condition for chemical equilibrium is that the chemical potentials of the components are equal in the phases that are in equilibrium. In the beginning, the α-phase with composition α1 is

4.1 Thermodynamics

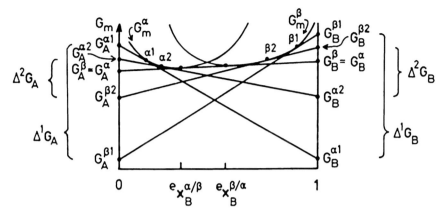

Fig. 4.6 Calculation of a two-phase equilibrium by moving the atoms in the direction of decreasing chemical potential (partial molar Gibbs energy G_i^ϕ) [6]

contacted with β-phase with a composition $\beta 1$. As seen from Fig. 4.6 (at the moment in question) the chemical potential (partial molar Gibbs energy) of component A in α-phase is $G_A^{\alpha 1}$ and in β-phase $G_A^{\beta 1}$ whereas that of component B in α-phase is $G_B^{\alpha 1}$ and in β-phase $G_B^{\beta 1}$, which are hardly equal. Thus, there is a driving force $\Delta^1 G_B \left(G_B^{\alpha 1} - G_B^{\beta 1} \right)$ driving the diffusion of B atoms to α-phase (from composition $\beta 1$ to $\alpha 1$) and $\Delta^1 G_A \left(G_A^{\beta 1} - G_A^{\alpha 1} \right)$ driving A atoms in the opposite direction. As the diffusion proceeds, the driving force for diffusion gradually decreases ($\Delta^2 G_B$ and $\Delta^2 G_A$) and vanishes when the chemical potentials of the components (A and B) become equal in both phases. This takes place when the two Gibbs energy curves for α- and β-phases have a common tangent and the equilibrium has been established.

4.1.5.3 Binary and Ternary Phase Diagrams

In order to obtain phase diagrams from the thermodynamic properties of the system, free energy composition diagrams together with the common tangent construction can be utilised. One such example is shown in Fig. 4.7. At temperature T_1 the only stable phase is liquid as temperature is above the melting points of elements A and B. At T_2 the system is below the melting point of A and is at the peritectic temperature where α and liquid react to form a new solid phase β. As the temperature decreases further to T_4 the γ-phase also becomes stable, and at T_5, the eutectic reaction ($l \rightarrow \beta + \gamma$) occurs in the system.

Binary alloys comprise two elements. In Fig. 4.8 an example of a binary composition temperature diagram, i.e. Pb–Sn binary phase diagram, is shown. The melting points of the constituting elements (Pb and Sn) have been marked on

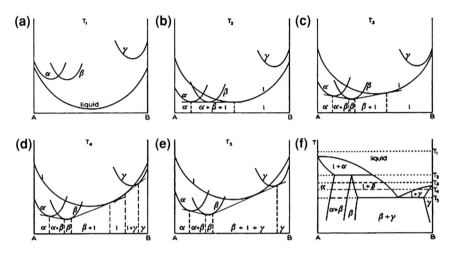

Fig. 4.7 Derivation of phase diagram form the free energy composition curves. $T_1 > T_2 > T_3 > T_4 > T_5$ [11]

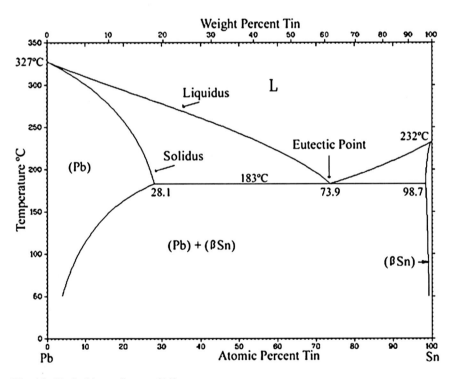

Fig. 4.8 Pb–Sn binary diagram [14]

4.1 Thermodynamics

the diagram (327°C for Pb and 232°C for Sn). Liquidus and solidus lines have also been marked in Fig. 4.8. Above the liquidus curve only the liquid phase is stable and below the solidus curve only the solid phases are stable. Between the liquidus and solidus there is an area, where both liquid and solid phases are stable. This is called a two-phase region. Solidus and liquidus temperatures can be determined for each nominal composition. Similarly, for a given nominal composition inside a two-phase region, the composition of the phases in equilibrium at a given temperature can be determined from the endpoints of the corresponding tie-line and their relative amounts by the lever rule (Eq. 4.34). In a specific (eutectic) composition the liquid phase is transformed into solid phases. In Fig. 4.9 the eutectic reaction L → (Pb) + (βSn) occurs at the eutectic point at 183°C. As the phase rule presented in Sect. 4.1.4 states, the reaction occurs at a certain temperature (the eutectic line is horizontal) and the composition of each phase, i.e. liquid, (Pb) and (β-Sn), is fixed. Only after the eutectic reaction has been completed can one of the degrees of freedom obtained, for example, change the temperature. On the left side of the diagram there is a region of terminal solid solution of fcc-(Pb) and on the right side a region of the terminal solid solution of (βSn). Tin has a maximum solubility to (Pb) of 28.1 atomic percent (at.%) and lead to (βSn) of 1.3 at.%. Other important reactions that may occur in phase diagrams in addition to eutectic reaction are peritectic (see Fig. 4.7) and monotectic reactions. Their solid state counterparts, eutectoid, peritectoid and monotectoid reactions also frequently occur in alloy systems. It must be emphasised here that one cannot obtain any information about kinetics or the morphology of the phase mixture from the phase diagram. The diagram only gives information about the phases that can be in equilibrium under certain composition-temperature combinations.

In a ternary system (Fig. 4.9a), where three elements are mixed, phase diagrams takes the standard form of a prism, which combines an equilateral triangular base (ABC) with three binary system "walls" (A–B, B–C, C–A). This three-dimensional form allows the three independent variables to be specified (two component concentrations and temperature). In practice, determining different sections of the diagram from these kinds of graphical models are difficult and therefore horizontal (isothermal) sections through the prism are used (Fig. 4.9b). The isothermal section is a triangle at a given temperature, where each corner represents the pure element, each side represents relevant binary systems and areas of different phases can be determined inside the triangle. Figure 4.9c shows a so-called Rozeboom diagram. In a Rozeboom diagram the liquidus surfaces are projected into the Gibbs triangle. Figure 4.9c also nicely exemplifies how a ternary system is constructed from relevant binary systems. In Fig. 4.10 an example of a space model of a ternary phase diagram is shown. From such a diagram it is relatively difficult to obtain useful data and therefore, as discussed above, isothermal sections are commonly taken from the space model and used to investigate phase equilibria at the particular temperature. In Fig. 4.11 a series of isothermal sections taken from the phase diagram shown in Fig. 4.10 are presented. At T_1 the temperature is below the melting points of B and the intermediate phase δ, but above the melting points of the other elements A and C. Thus, one can see the two solid one-phase

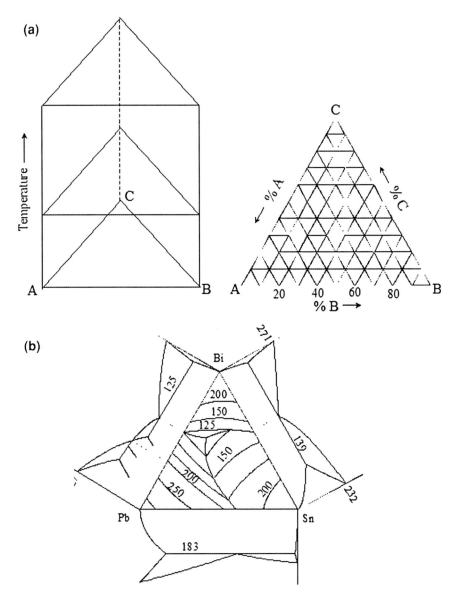

Fig. 4.9 **a** Ternary system, **b** isothermal section at $T = x°C$ and **c** Rozeboom diagram of the Bi–Pb-Sn system

areas (β and δ) and the two-phase regions (L + β) and (L + δ) as well as the large liquid phase area. At temperature T_2 one can see the first three-phase equilibrium

4.1 Thermodynamics

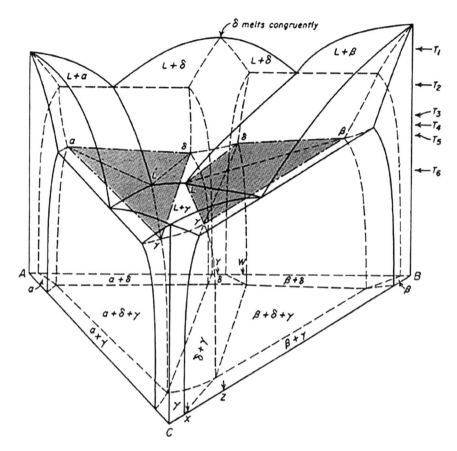

Fig. 4.10 Space diagram of ternary phase diagram with two eutectic systems and one system where there is a congruently melting intermediate phase. [15]

to appear (L + β + δ). At this temperature all three end-members are also solid (see the single α, β, and γ -phase regions near the corners). As the temperature further decreases more three phase equilibria appear and the two binary eutectic reactions take place and finally at the temperature T_5 the last liquid solidifies.

In addition to the isothermal section vertical sections (isopleths) can also be taken from the space diagram of a given ternary system (Fig. 4.12). Although the isopleths appear quite like binary phase diagrams they must not be confused with them. In general tie-lines cannot be used with isopleths and they only show the temperature composition regions of the different phases. More information about the utilisation of different sections and higher order phase diagrams can be found, for example, from [11, 12, 15].

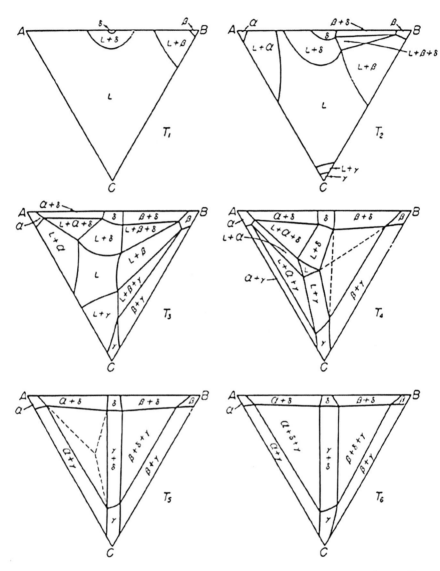

Fig. 4.11 Series of isothermal sections from the space diagram shown in Fig. 4.10 ($T_1 > T_2 > T_3 > T_4 > T_5 > T_6$) [15]

4.1.6 Calculation of Phase Diagrams

Phase diagrams are graphical representations of the domains of stability of the various classes of structures (one phase, two phase, three phase etc.) that may exist in a system at equilibrium. Phase diagrams are most commonly presented in the temperature –pressure–composition space. In the context of this chapter, only

4.1 Thermodynamics

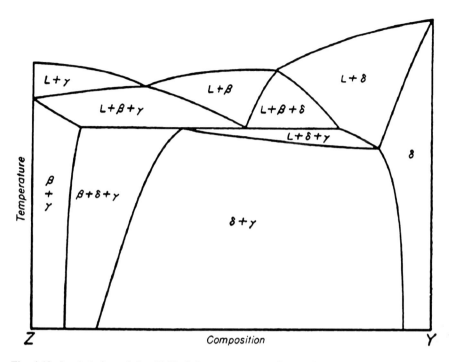

Fig. 4.12 Isopleth through line Z–Y of the ternary space diagram in Fig. 4.10 [15]

systems where the pressure variables can be neglected are considered (as the binary A–B phase diagram in Fig. 4.1a and in the ternary systems shown in the previous chapter). Many other coordinate systems, for example activity of one component versus mole ratio of the others, are also possible [16].

To represent the phase equilibria in ternary system at constant pressure, a three-dimensional construction is required, as shown above. However, as many practical processes are carried out at constant temperature, the most quantitative ternary phase diagrams are presented as isothermal sections at certain temperatures. The most common method for plotting composition in a ternary system uses an equilateral triangle, sometimes referred to as the Gibbs triangle introduced in Sect. 4.1.5. As an example, the ternary Si–Ta–Cu phase diagram used to study phase relations in the Si/Ta/Cu metallization system [17, 18] is shown in Fig. 4.13. At the corners are the pure elements Ta, Si and Cu and the three edges representing the Cu–Si, Ta–Si and Cu–Ta binary systems. There are no ternary phases in this system. Triangles in the diagram, bounded by three straight lines, represent three-phase equilibrium, such as Cu + $TaSi_2$ + Ta_5Si_3 in Fig. 4.13. Two tie-lines cannot cross each other, since at the point of intersection there would be four phases in equilibrium and this would violate the Gibbs phase rule (at a chosen temperature: $f = c(= 3) - p(= 4) = -1$).

Phase diagrams have sometimes been regarded as something that can only be determined experimentally. However, as a phase diagram is the manifestation of

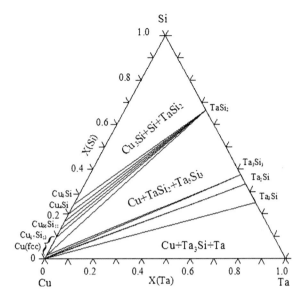

Fig. 4.13 Isothermal section at 973 K from the evaluated ternary Si–Ta–Cu phase diagram

the state of equilibrium, it is possible to construct any kind of phase diagram if the equilibrium state of the system has been calculated. This in turn requires the evaluation of the thermodynamic properties of the corresponding system by assessing all the available experimental information in thermodynamic terms. Generally, one is interested in equilibria under constant pressure and therefore the Gibbs free energy is the expedient thermodynamic function. Analytical expressions for the free energy functions of all phases must be derived first. The thermodynamic models used in the description of the Gibbs free energy of different phases are important, since successful and reliable calculation relies on the appropriate choice of model for each phase appearing in the system. Then by summing up all the Gibbs free energies of individual phases, the phase equilibria can be computed by minimising the total Gibbs free energy of the system. The mathematical expressions for the Gibbs free energy of the individual phases contain parameters which have to be optimised to give the best fit to all the experimental information available. A major difficulty arises from the fact that the value of a parameter (which is used in the description of a simple system) will affect the evaluation of all the related higher systems. Thus, one should use as much information as possible from different sources in each optimisation process. The preceding approach is known as the CALculation of PHAse Diagrams (CALPHAD) method [18–20]. The CALPHAD method is based on the axiom that complete Gibbs free energy versus composition curves can be constructed for all structures exhibited by the elements right across the whole alloy system. This involves the extrapolation of (G, x)-curves of many phases into regions where they are metastable (see Fig. 4.1b). This requires that the relative Gibbs free energies for various crystal structures of the elements of the system must be established. These are known as lattice stabilities and the Gibbs free energy differences

4.1 Thermodynamics

between all the various potential crystal structures in which an element can exist need to be characterised as a function of temperature, pressure and volume [20–22]. As an example of a simple binary phase diagram, the calculation based on the regular solution model is shown in Fig. 4.14. It demonstrates nicely how by varying the interaction parameters in α and β phases most of the typical binary phase diagram features can be reproduced. Typically, however, the mathematical models of the phases are more complicated.

CALPHAD method is commonly used for evaluating and assessing phase diagrams. The power of the method is clearly manifested in its capability to extrapolate higher order systems from lower order systems, which have been critically assessed, thus reducing the number of experiments required to establish the phase diagram. The determination of binary equilibrium diagrams usually involves the characterisation of only a few phases, and experimental thermodynamic data on each of the phases is generally available in various thermodynamic data banks as well as in the literature. However, when handling multicomponent systems or/and metastable conditions there is a need to evaluate the Gibbs free energies of many phases, some of which may be metastable over much of the composition space. In addition, the need for reliable data grows rapidly when multicomponent systems are considered. For example to establish a diagram for a five-component system one needs carefully assessed data on ten binary systems, ten ternary systems and five quaternary systems. Readers interested in the actual thermodynamic modelling procedures and issues and problems associated with them are referred to the vast amount of available literature, for example review articles and books [20–24]. It should finally be emphasised that it is not possible to determine a ternary phase diagram solely based on the data from binary phase diagrams, since the binary data does not yield information about ternary interaction parameters. Therefore, experimental work is always required in the critical assessment of phase diagrams as discussed above.

Phase diagrams are useful for concisely presenting the equilibrium relationships in alloy systems. However, since phase diagrams are equilibrium diagrams they do not provide any information about reaction rates, the effect of defects or the phase distribution and morphology. Nevertheless, because the equilibrium diagrams are directly related to free energy versus composition diagrams, information about the driving forces of different reactions in the system can be obtained from the (G, x)-diagrams. It is also possible to describe metastable phases with the help of (G, x)-diagrams and establish relative stabilities of the phases.

In order to take into account reaction rates, among other things, one needs to bring kinetics into the analysis. Therefore, the thermodynamic data of the system must be combined with the available kinetic data to provide information about time dependence and reaction sequences during phase transformations. This requires the assumption of local equilibrium at the interfaces.

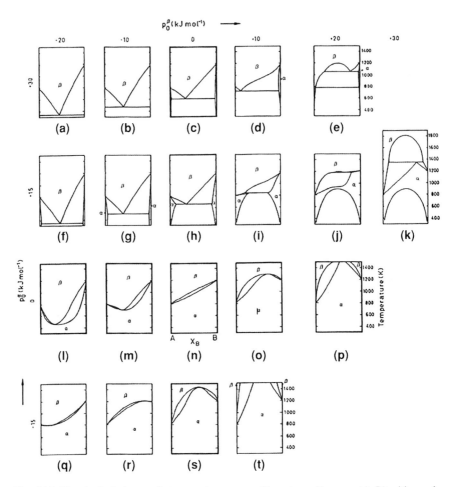

Fig. 4.14 Topological changes in temperature-composition phase diagrams (A–B) with regular phases α and β brought about by systematic changes in the regular solution parameters p_o^α and p_o^β (Entropies of transformation of pure A and B taken as 10.0 J/mol) [16]

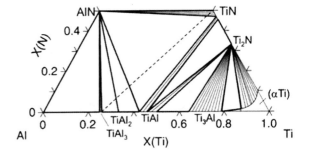

Fig. 4.15 Calculated isothermal section of the Ti-Al-N phase diagram at 600°C, with the ternary compounds suspended (redrawn from [32]). The *dashed line* represents the TiAl$_3$–TiN equilibrium suggested in [33]

4.1.7 Local Equilibrium in Thin Film System?

Although local equilibrium can be assumed in a bulk solid-state diffusion couple and this assumption has been taken as the theoretical basis for phase diagram determination by using diffusion couples [25, 26] it is well established that in planar binary thin-film diffusion couples not all of the compound phases predicted by the equilibrium phase diagram can be observed to be present simultaneously. In the case of silicides, for example, it is even common that only one compound in the equilibrium phase diagram occurs between Si and metal thin films, such as in the Ni–Si or Cu–Si system [27]. Other interesting examples are the cases with the bulk Ti/Al and Cu/Si diffusion couples where a well adherent layer of $TiAl_3$ and Cu_3Si formed at the interface, but no other intermetallic compounds showed up in either of the couples.

Furthermore, it is well known that the first phase to form through an interface reaction need not be the most stable phase but in fact is often the least thermodynamically favoured phase according to the Ostwald's rule, as indicated above. A prominent example is the formation of amorphous phases at interfaces in diffusion couples. In soldered joints, although the reaction products between Cu and Sn or Sn-bearing solders are exactly those predicted from the equilibrium phase diagram, i.e. ε-Cu_3Sn and η-Cu_6Sn_5 from the pure copper side [28–30], a high temperature metastable phase Cu_4Sn has been detected in a thermally annealed Cu/Sn thin film system [31]. The situation would become more complicated in multicomponent systems as have been seen in the Ti–Al–N system [32]. It was experimentally observed that, in the multilayer system Al/Ti/TiN, where the thickness ratio of the Al layer to Ti layer was 3, the pure metal layers Al and Ti were completely converted into $TiAl_3$ after annealing at 600°C for 10 min. No additional interdiffusion occurred in this in situ formed $TiAl_3$/TiN diffusion couple after it was further annealed at 600°C for 20 h. Based on these results, equilibrium was proposed to exist between $TiAl_3$ and TiN [33]. However, thermodynamic calculation shows that the tie-line $TiAl_3$–TiN is not stable at this temperature (Fig. 4.15) The reason for this discrepancy is that the diffusion flux through $TiAl_3$ is several orders of magnitude higher than those through the other Ti–Al binary compounds and hence $TiAl_3$ is kinetically very stable [34]. Thus, in order to complete the picture about interfacial reactions, kinetic considerations must be taken into account and a combined thermodynamic-kinetic method used.

4.1.8 The Use of Gibbs Energy Diagrams

Figure 4.7 shows how the molar Gibbs free energy diagrams are directly connected to T–x_i equilibrium diagrams. The Gibbs free energy diagrams provide also other important information and can be used to obtain data about driving forces for diffusion or chemical reaction, grain boundary segregation and so on. Thus in this chapter some of the basic properties of binary molar property diagrams are reviewed (all issues discussed here apply also to ternary cases).

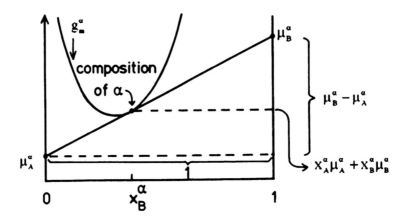

Fig. 4.16 Evaluation of molar Gibbs energy g_m of a binary mixture from the chemical potentials of A and B [6]

In Fig. 4.16 a typical molar property diagram at a given temperature T_0 is shown. The diagram shows some of the basic properties of a molar property diagram. From Fig. 4.16, one can see, for example, how a Gibbs energy of a phase is defined $(g_m^\alpha = x_A^\alpha \mu_A^\alpha + x_B^\alpha \mu_B^\alpha)$ with the help of chemical potentials, how the chemical potentials for a species in a given phase are defined (from the endpoints of the tangent), so that the slope of a tangent equals the chemical potential difference of B and A in the alpha—phase $\left(\frac{dG_m^\alpha}{dx_B} = \mu_B^\alpha - \mu_A^\alpha\right)$ and so on. These simple geometrical features of the molar diagrams can be used for a wide variety of applications [6].

In Fig. 4.17 the common tangent construction defined above is used to determine the driving forces for diffusion of Cu through Cu_3Sn and Sn through Cu_6Sn_5. $\Delta_r G_{Cu}^0(\varepsilon)$ is the chemical potential difference of Cu between interfaces Cu_3Sn/Cu and Cu_6Sn_5/Cu_3Sn, which drives the diffusion of Cu through Cu_3Sn and $\Delta_r G_{Sn}^0(\eta)$ is the chemical potential difference of Sn between the interfaces Sn/Cu_6Sn_5 and Cu_6Sn_5/Cu_3Sn, which drives the diffusion of Sn through the Cu_6Sn_5 layer. It is easy to realise from Fig. 4.17 that changes in the stabilities of η and ε phases (in their Gibbs free energy) will change the values of $\Delta_r G_{Sn}^0(\eta)$ and $\Delta_r G_{Cu}^0(\varepsilon)$ and thus increase or decrease the driving forces for diffusion of components in the system. Examples of this will be shown in Chap. 6. From Fig. 4.17 other important features can also be extracted. For instance, the common tangent between the curves g_{bin}^L and g^η gives the equlibrium solubility of Cu to a liquid solder (L), which is given as $^e x_{Cu}^L$. The equilibrium solubility is the amount of Cu that can be dissolved infinitely slowly to liquid solder before the η-phase comes into equilibrium with the liquid. The formation of η-phase does not, however, occur with this composition, as the driving force is zero at this point (no supersaturation). In real cases the dissolution of Cu does not take place infinitely slow and, therefore, the

4.1 Thermodynamics

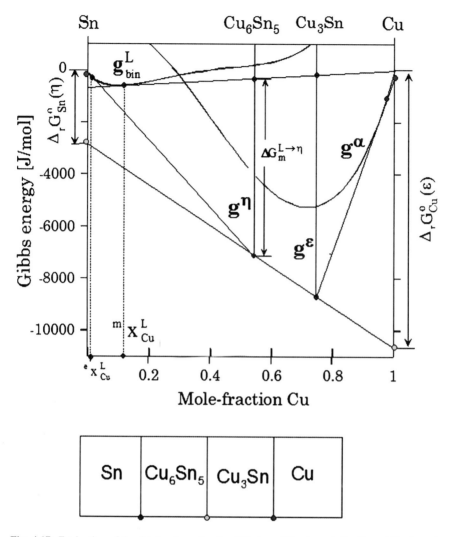

Fig. 4.17 Evaluation of the driving force for the diffusion of Cu through Cu$_3$Sn and Sn through Cu$_6$Sn$_5$ at 235°C [35]

equilibrium solubility is generally exceeded. The solubility of Cu does not increase infinitely, but there is an upper limit for its value and this can also be determined from Fig. 4.17. When more and more Cu dissolves into liquid eventually a situation is faced where dissolution of more Cu would lead to precipitation of pure metallic Cu out of the supersaturated solder. This corresponds to the common tangent construction between the solder and pure Cu. The tangent point in the liquid curve at this metastable equilibrium gives the upper value of Cu that can be dissolved into liquid solder at any rate, i.e. the metastable solubility $^m x^L_{Cu}$. When this value has been reached, the driving force for the formation of η-phase has also

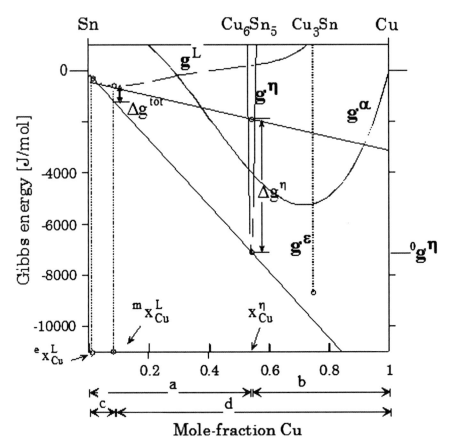

Fig. 4.18 Evaluation of the driving force for the formation of Cu$_6$Sn$_5$ as well as the change in Gibbs energy of the system when η phase is precipitated out from the supersaturated solder at 235°C [35]

reached its maximum (shown in the diagram as $\Delta G_m^{L \to \eta}$ in Fig. 4.17). Since the metastable solubility is directly related to the dissolution rate of a given metal to a solder in question, it provides important information about the formation of intermetallic compounds between different metals and solders.

One can also obtain quantitative numerical data about energy changes in a given system, by utilising the simple geometric constructions shown above together with some simplified assumptions. Let us take one example from the important Cu–Sn system (Fig. 4.18). Consider the nucleation and growth of the η-phase, with $x_{Cu}^\eta = a$ and $x_{Sn}^\eta = b(a+b=1)$, out of supersaturated solder $^m x_{Cu}^L = c$. The driving force for the nucleation of the η-phase is given by $\Delta g^\eta = g_{tg}^L - ^e g_{tg}^L$ (per mole of η). The two terms can be expanded as:

$$g_{tg}^L = x_{Cu}^\eta \mu_{Cu}^L (^m x_{Cu}^L) + (1 - x_{Cu}^\eta) \mu_{Sn}^L (^m x_{Cu}^L) \tag{4.37}$$

4.1 Thermodynamics

And

$$^e g^L_{tg} = x^{\eta}_{Cu} \mu^L_{Cu}(^e x^L_{Cu}) + (1 - x^{\eta}_{Cu}) \mu^L_{Sn}(^e x^L_{Cu}) \tag{4.38}$$

which gives:

$$\Rightarrow \Delta g^{\eta} = x^{\eta}_{Cu} \mu^L_{Cu}(^m x^L_{Cu}) + (1 - x^{\eta}_{Cu}) \mu^L_{Sn}(^m x^L_{Cu}) \\ - [x^{\eta}_{Cu} \mu^L_{Sn}(^e x^L_{Cu}) + (1 - x^{\eta}_{Cu}) \mu^L_{Sn}(^e x^L_{Cu})] \tag{4.39}$$

and by using the definition of the chemical potential of a component this gives:

$$= RT \left[x^{\eta}_{Cu} \ln \left(\frac{^m a^L_{Cu}}{^e a^L_{Cu}} \right) + x^{\eta}_{Su} \ln \left(\frac{^m a^L_{Su}}{^e a^L_{Su}} \right) \right] \tag{4.40}$$

If we simplify the treatment by assuming that the liquid behaves as a perfect solution, the activity values can be replaced by compositions, which give:

$$\cong RT \left[x^{\eta}_{Cu} \ln \left(\frac{^m x^L_{Cu}}{^e x^L_{Cu}} \right) + (1 - x^{\eta}_{Cu}) \ln \left(\frac{1 - ^m x^L_{Cu}}{1 - ^e x^L_{Cu}} \right) \right]$$

$$\cong 8.3145 \, \text{J/Kmol} \times 508 \, \text{K} \left[0.545 \ln \left(\frac{0.06}{0.018} \right) + (1 - 0.545) \ln \left(\frac{1 - 0.06}{1 - 0.018} \right) \right]$$

$$= 2687.5 \, \text{J/Kmol} \tag{4.41}$$

Similarly, the change in Gibbs energy of the system owing to the precipitation of Cu_6Sn_5 can be expressed as:

$$\Delta g^{tot} = g^L_{tg} - ^e g^L_{tg} (= \Delta g^{L \to L + \eta}) \tag{4.42}$$

This can be written as:

$$= RT \left[^m x^L_{Cu} \ln \left(\frac{^m a^L_{Cu}}{^e a^L_{Cu}} \right) + ^m x^L_{Sn} \ln \left(\frac{^m a^L_{Sn}}{^e a^L_{Sn}} \right) \right] \tag{4.43}$$

and by again assuming perfect behaviour:

$$\cong RT \left[^m x^L_{Cu} \ln \left(\frac{^m x^L_{Cu}}{^e x^L_{Cu}} \right) + (1 - ^m x^L_{Cu}) \ln \left(\frac{1 - ^m x^L_{Cu}}{1 - ^e x^L_{Cu}} \right) \right]$$

$$\cong 8.3145 \, \text{J/Kmol} \times 508 \, \text{K} \left[\begin{array}{c} 0.06 \ln \left(\dfrac{0.06}{0.018} \right) \\ + (1 - 0.06) \ln \left(\dfrac{1 - 0.06}{1 - 0.018} \right) \end{array} \right] \tag{4.44}$$

$$= 131.6 \, \text{J/Kmol}$$

Hence, it can be concluded that these simple molar diagrams (Gibbs free energy diagrams) give an extensive amount of important information in an easily visualised form.

4.2 Diffusion Kinetics

Thermodynamics provides the basis for analysing reactions between different materials. However, one cannot predict the time frame of the reactions on the basis of the phase diagrams. Therefore, diffusion kinetics must be brought into the analysis. In this section the basics and some of the related nomenclature of solid-state diffusion are introduced. Several useful relations between different types of diffusion coefficients are also introduced.

4.2.1 Multiphase Diffusion

The driving force for interdiffusion is the difference in concentration (or more precisely in chemical potential or activity). Typically, diffusion kinetic analyses are based on the standard solutions of Fick's laws under certain boundary conditions. In the case of the steady state materials flow, i.e. when the concentration remains constant with time, Fick's first law states that the flux of element i (J_i) is directly proportional to the product of the diffusion coefficient (D) and the concentration gradient of an element i [36]:

$$\frac{1}{A}\frac{dm}{dt} = J = -D\frac{\partial c}{\partial x} \quad (4.45)$$

For non-steady state diffusion, in which the concentration varies with distance and time, Fick's second law is applied [36]:

$$\frac{\partial c}{\partial t} = -\frac{\partial J}{\partial x} = \frac{\partial}{\partial x}\left(D\frac{\partial c}{\partial t}\right) \quad (4.46)$$

Thus, at constant temperature the rate of change of concentration is equal to the diffusivity times the rate of change of the concentration gradient. The utilisation of Eq. 4.18 requires there to be a significant (measurable) composition difference within a phase. Then the diffusion coefficients can be obtained by using either Bolzmann-Matano or Sauer-Freise analysis [37]. However, if the system includes phases that have narrow homogeneity ranges (like many intermetallic compounds in solder interconnections), the above-mentioned analyses are not applicable and the method derived by Wagner [38] needs to be used.

Utilization of Fick's law requires that there are no other driving forces than concentration gradient. If other forces are present one must add the so-called drift term to the flux equation:

$$J = -D\frac{\partial c}{\partial x} + \langle v \rangle c \quad (4.47)$$

For the drift velocity there is a Nernst–Einstein relation that states:

4.2 Diffusion Kinetics

$$v_D = \frac{DF}{kT} \qquad (4.48)$$

where F is the driving force. Several types of gradients can drive diffusion. For example, if the additional driving force is electric field (as in electromigration):

$$F = qE \qquad (4.49)$$

The driving force (F) can be further defined as:

$$F_{EM} = Z^* e \vec{E} \qquad (4.50)$$

where Z^* is a dimensionless quantity called the effective valence taking both electron wind and valency effects into account, E is the electric field [V/cm] and e is the electron charge. For conducting metal:

$$F_{EM} = Z^* e \rho j \qquad (4.51)$$

where ρ is the resistivity of the metal in question and j is current density.

One then obtains the flux due to electromigration as:

$$J_i = \frac{Dc_i}{kT} Z^* e \rho j \qquad (4.52)$$

In many cases the additional driving force is chemical in nature. The total Gibbs energy of the system can be expressed as a function of chemical potentials, which is related to the activities of the components as follows [39]:

$$G_i^\phi \equiv G_i^0(T) + RT \ln a_i. \qquad (4.53)$$

Since the chemical potential of a component i will have the same value in all the equilibrated phases, the difference in activity of a component (i.e. the driving force for its diffusion) will vanish at equilibrium. A fundamental condition is that no atom can diffuse intrinsically against its own activity gradient [40]. If a maximum is found experimentally in the activity profile of a component it is caused by the intrinsic movement of the other components. This condition is used as a criteria when thermodynamic prerequisites for the diffusion of certain species in a given reaction sequence is rationalised with the help of activity diagrams. In addition, the thermodynamic data obtained from the simulations can be utilised also in quantitative diffusion flux calculations as shown next.

The flux of a component A in, for instance, an A–B binary alloy can be written in terms of the gradient in chemical potential μ (the driving force F in the Nernst–Einstein relation):

$$J_A = -B_A C_A \frac{\partial \mu_A}{\partial x}, \qquad (4.54)$$

where B_A is called the atomic mobility of A. The chemical potential of A, μ_A, can be written in terms of activity, a_A:

$$\mu_A = \mu_A^\theta + RT \ln a_A, \quad (4.55)$$

where μ_A^θ is the chemical potential at the chosen standard state ($T = 298$ K; $p^\theta = 1$ bar) in J/mol and R is the gas constant (8.3143 J/mol K). Substituting Eq. 4.54 in Eq. 4.53 and knowing that $dN_A = (V_m^2/V_B)dC_A s$ [41] and $C_A = N_A/V_m$ one can obtain:

$$J_A = -B_A RT \frac{V_m}{V_B} \left(\frac{\partial \ln a_A}{\partial \ln N_A}\right) \frac{dC_A}{dx}, \quad (4.56)$$

With $\frac{d \ln a_A}{d \ln N_A} = \frac{d \ln a_B}{d \ln N_B}$ being the *thermodynamic factor* (a_A and a_B are the chemical activities of components and mole fraction is now denoted by N_i). The standard state is generally the pure element (Raoultian) reference state.

The mobility of A, B_A, can be experimentally determined from a tracer experiment. If one deposits an "infinitely" thin layer of a radioactive isotope of A on the surface of a homogeneous A–B alloy, the activity as a function of the distance in the diffusion direction after annealing can be measured using the sectioning technique. By applying the method of Gruzin [42], in which it is assumed that the atoms A and isotopes A* move identically, a so-called *tracer diffusion coefficient* D_A^* can be derived. The Nernst–Einstein relation directly connects the tracer diffusion coefficient to the atomic mobility:

$$D_i^* = B_i RT. \quad (4.57)$$

When we now take Fick's law, Eqs. 4.56 and 4.57, we arrive at a general relation between the intrinsic and tracer diffusion coefficients:

$$D_A = D_A^* \frac{V_m}{V_B} \left(\frac{\partial \ln a_A}{\partial \ln N_A}\right). \quad (4.58)$$

If the pertinent thermodynamic data are available, Eq. 4.58 allows the calculation of intrinsic diffusivities from tracer diffusion coefficients, which will greatly reduce the amount of experimental work necessary when phase growth behaviour in specific system is investigated. It is to be noted here for example that when using marker experiments to determine intrinsic diffusion coefficients (D_i's) (needed in the thermodynamic-kinetic analysis) they can be only determined at the Kirkendall-plane (i.e. at one composition). Thus, for a phase with wide solubility range many diffusion couples and/or multifoil experiments should be carried out to determine the D_i's over the appropriate concentration range. However, when thermodynamic assessment of the system is at hand the D_i's can be calculated with the help of Eq. 4.58.

It is often the case that the phase under investigation grows with a very narrow homogeneity range following the phase diagram and it is not possible to determine the vanishingly small concentration gradient to calculate the interdiffusion coefficient. To circumvent this problem, Wagner [38] introduced the concept of the

4.2 Diffusion Kinetics

integrated diffusion coefficient $(\tilde{D}_{\text{int}})$ and can be expressed in terms of composition of i as

$$\tilde{D}_{\text{int}} = \int_{N'v}^{N''v} \tilde{D} dN_i = \tilde{D} \Delta N_i \qquad (4.59)$$

where $\tilde{D} (m^2/s)$ is the interdiffusion coefficient, N_i is the composition of i and $\Delta N_i = N_i'' - N_i'$ is the narrow homogeneity range of the product phase.

Furthermore, interdiffusion coefficient, $\tilde{D}$ is related to the tracer diffusion coefficient of the species, D_A^* and D_B^* by

$$\tilde{D} = \left(N_A D_B^* + N_B D_A^* \right) W \left(\frac{d \ln a_B}{d \ln N_B} \right) \qquad (4.60)$$

where N_A and N_B are the compositions of A and B in the AB-phase, $\frac{d \ln a_B}{d \ln N_B}$ is the thermodynamic parameter, a_B is the activity of the element B and W is the vacancy wind effect and can be in most cases considered as one [40]. By replacing Eq. 4.60 in Eq. 4.59 and following standard utilisation of common tangent construction in Gibbs energy diagrams [8, 21]

$$\tilde{D}_{\text{int}} = \int_I^{II} \left(N_A D_B^* + N_B D_A^* \right) N_B \, d \ln a_B = \left(N_A D_B^* + N_B D_A^* \right) N_B \left[\ln a_B^{II} - \ln a_B^I \right]$$

$$(4.61)$$

where I and II represent the interfaces from which the product phase grows.

Moreover, following Eq. 4.61 and from the standard thermodynamic relation, $\mu_B = G_B + RT \ln a_B$, we can write as:

$$\tilde{D}_{\text{int}} = \left(N_A D_B^* + N_B D_A^* \right) \frac{N_B \left(\mu_B^{II} - \mu_B^I \right)}{RT} = -\left(N_A D_B^* + N_B D_A^* \right) \frac{N_B \Delta_r G_B^o}{RT} \qquad (4.62)$$

where G_B is the free energy of pure B, μ_B is the chemical potential of B, and $\Delta_r G_B^o$ is the driving force for the diffusion of B. The values of $\Delta_r G_B^o$ under different ambient conditions can be calculated from the appropriate data.

An important guideline to remember is that the penetration depth of a given species by diffusion is given as $\sqrt{Dt}$. This simple relation is very indicative when scales of different kinetic phenomena are considered. For example, with diffusion coefficients 10^{-5} cm^2/s (a typical value for diffusion in liquid), 10^{-8} cm^2/s (a typical value for a cubic solid metal at its melting point) and 10^{-15} cm^2/s (a typical value for bulk diffusion in solid state) the penetration depths during 1 s are the order of 10^{-4}, 10^{-5} and 10^{-8} m, respectively. As the time for diffusion is proportional to x^2/D with the same values for diffusion coefficients as above, the

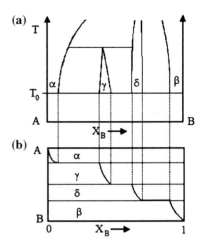

Fig. 4.19 Schematic binary phase diagram and the corresponding reaction structure at T_0. (redrawn from Ref. [39])

time required for movement of 1 μm are, 10^{-3}, 1 and 10^6 s (about 116 days), respectively.

4.2.2 Diffusion Couples and the Diffusion Path

Generally, in a binary system the interaction between components and the resulting reaction products can be predicted by using the relevant phase diagram. After annealing at a certain temperature for a sufficiently long time, all the thermodynamically stable phases of the system at that particular temperature (and only them), will exist as layers between the end members. According to the phase rule only single-phase homogeneous regions can be formed, the interfaces between various phases must be macroscopically planar, and the concentrations at the interfaces of various phases can be read from the phase diagram (Fig. 4.19).

If certain phases are absent, it may be caused by nucleation problems or by the "kinetic" stability described briefly in Sect. 4.1.7. In cases when there are no nucleation problems associated with the phase formation and the diffusion is the rate-limiting step the phase sequence after the reaction is determined by the corresponding phase diagram. In principle, in a specific binary system one can also say something about the temporal evolution of the system, if the interdiffusion coefficients in all the appearing phases are known. However, when one is dealing with higher order systems the prediction of the reaction layers becomes much more difficult. In a ternary system the additional degree of freedom enables also the formation of two-phase regions and curved interfaces. This means that theoretically many different phase sequences are possible and the final structure cannot be, in general, interpreted directly from the phase diagram. In addition, the concentration and activity gradients are not necessarily in the same direction as generally in the binary systems, thus enabling, for example, the so-called "up-hill" diffusion.

4.2 Diffusion Kinetics

Phase transformation will be diffusion controlled when the mass transport through the bulk is the rate-limiting step. Then the boundary conditions governing the rate of diffusion can be evaluated by assuming that whenever two phases meet at the interface, their compositions right at the interface are very close to those required by the equilibrium, and the thermodynamic potentials are continuous across the interface. This is called the local equilibrium approximation [21]. If the local equilibrium can be assumed in the system, it is often possible to use phase diagrams coupled with a few rules to predict possible or at least to rule out impossible reaction sequences in ternary systems. The first rule is concerned with the principle of mass-balance, which requires that material cannot be created or destroyed during reaction. This requires that the diffusion path, which is a line in the ternary isotherm, representing the locus of the average compositions parallel to the original interface through the diffusion zone [41–43], crosses the straight line connecting the end members of the diffusion couple at least once. During the reaction, the system will follow only one unique, reproducible diffusion path. Kirkaldy et al. [44–46] have presented a number of rules which the diffusion path must obey.

The concept of diffusion path can be comprehended with the help of Fig. 4.20, which shows the hypothetical A–B–C phase diagram with two binary compounds X and Z and the ternary compound T [40]. In Fig. 4.20 four diffusion paths are given with the corresponding morphologies of the diffusion zone, for the diffusion couple C versus X. They all fulfil the mass-balance requirement. This also gives constraints about the relative thicknesses of the various diffusion layers like T and Z in Fig. 4.20b. The Z-phase has to be much thinner than the T-phase in order to comply with the mass balance. This is because the Z-phase is much further from the contact line (e.g. average composition) than the T-phase. Not only the relative thicknesses, but also the total thickness of the diffusion layer is related to the diffusion path. If for instance in the Z-phase the diffusion is very slow and in the T-phase fast, then the total layer width for Fig. 4.20b or c will be smaller than in the case of Fig. 4.20d. In the first examples, the continuous Z-layer acts as a kind of diffusion barrier, whereas in the latter case the total layer thickness is governed by the faster diffusion in the T-phase, which is probably only slightly hindered by the discontinuous Z-particles. One must realise that diffusion path corresponds only to a topological distribution of phases in space.

It is difficult to predict the diffusion path or to exclude impossible ones with the help of the mass-balance requirement alone. Thus, van Loo et al. [40] have introduced another rule, which is based on the thermodynamic driving force behind the diffusion phenomenon. The rule states that no element can diffuse intrinsically against its own activity gradient. If this took place, it would mean that an atom should diffuse from a low chemical potential area to a high chemical potential area—a process that does not spontaneously occur in nature. By calculating the chemical activities of components as the function of the relative atomic fractions of the elements, and taking into account the above-mentioned mass-balance considerations, the sequence of compounds formed during reaction can often be predicted as can be seen from the following examples.

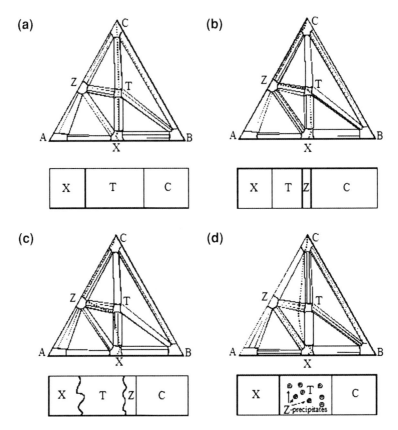

Fig. 4.20 Examples of possible morphologies for the reaction layer in a diffusion couple X/C. The corresponding diffusion paths are plotted on the isotherms (redrawn from [40])

An example from the literature is shown in Fig. 4.21, which is redrawn from [47]. It shows the isothermal section from the Cr–Si–C system at 1273 K. The experimentally determined diffusion path is superimposed into the isothermal section and into the activity diagram. The activity diagram shown on the right-hand side of Fig. 4.20 is one form of many different types of stability diagrams. In such a diagram the thermodynamic potential of one of the components is plotted as a function of the relative atomic fractions of the other two components. The activity values needed can be calculated from the assessed thermodynamic data. When calculating the activities of the components, the activities of the stoichiometric compounds at equilibrium are regarded to be one. It should be noted that the precision of the calculations is very much dependent on the accuracy and consistency of the thermodynamic data used. Therefore, great care should be exercised when using data from different sources.

In an activity diagrams the stoichiometric single-phase regions are represented as vertical lines, two-phase regions as areas and three phase fields as horizontal

4.2 Diffusion Kinetics

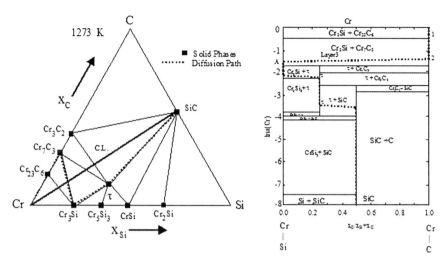

Fig. 4.21 Experimentally determined isothermal section and the calculated activity diagram at 1273 K. The contact line (C.L.) and the diffusion path are also superimposed onto both diagrams. τ is the ternary phase (redrawn from Ref. [47])

lines. The vertical left- and right-hand axes represent the binary edge systems. From Fig. 4.21 it is evident that Cr moves along its decreasing activity in the experimentally observed reaction sequence $SiC/\tau/Cr_3Si/Cr_3Si + Cr_7C_3/Cr_7C_3/Cr_{23}C_6/Cr$. The phase sequence $SiC/Cr_3C_2/\tau/CrSi/Cr_5Si_3/Cr_3Si/Cr$, another example of a diffusion path that fulfils the mass-balance requirement, is impossible from the thermodynamic point of view, since Cr would have to diffuse against its own activity gradient inside the ternary phase τ to get from CrSi to Cr_3C_2.

It should be noted that in thin film samples (as when working with very thin diffusion barriers) the assumption of local equilibrium, which is necessary for the above treatment, is not self-evident. The assumption of local equilibrium requires that reactions at the interfaces are fast enough so that all the atoms arriving in the reaction region are used immediately and the rate-determining step is diffusion, as discussed above. However, with very thin layers this requirement may not be fulfilled. The reasons for this originate mainly from special conditions prevailing during thin film reactions: (a) relatively low reaction temperatures, (b) small dimensions, (c) high density of short-circuit diffusion paths, (d) relatively large stresses incorporated in thin films, (e) relatively high concentration of impurities, (f) metastable structures, (g) large gradients, and so on. These conditions mean that the complete thermodynamic equilibrium is hardly ever met in thin film systems. However, the local equilibrium is generally attained at interfaces quite rapidly. Therefore, the procedure presented above provides a feasible analysis method also for studying thin film reactions. Finally, it must be emphasised that there is no fundamental difference between thin film and bulk diffusion couples, and the differences often observed (i.e. sequential versus simultaneous phase formation) occur because of the structural features of thin film structures (see the list above).

4.3 Microstructures

As the microstructure of a material determines many of its properties and their evolution under different environmental conditions, it is of great importance to understand the origin of the different microstructures. Thermodynamics and diffusion kinetics provide the basis for the formation of microstructures, as already discussed, but additional information is required to fully determine the formation and evolution of these features. Different aspects, such as grain boundaries, texture, defect type and density, segregation of impurities etc. all contribute to the final microstructure observed with the help of different microscopes. Thus, in the following, several aspects related to the above defined list will be discussed.

4.3.1 The Role of Grain Boundaries

Thin films possess usually high density of grain boundaries (Fig. 4.22), which can have a significant effect on the growth kinetics. This is because of the enhanced atom transport via the short-circuit paths. A simple situation readily occurs in thin film experiments: columnar grains, with their long axis along the direction of the diffusion flux. This situation can be modelled by dividing the film into two different parts: one with diffusion coefficient D_{vol} (lattice) and the other with diffusion coefficient D_{gb} (grain boundary). The number of atoms transported per unit area and unit time is given

$$\text{by: } M(t) = \left(A_l J_l + A_{gb} J_{gb}\right) = \left(A_l D_{vol} + A_{gb} D_{gb}\right) \frac{dc}{dx} \quad (4.63)$$

where A_l and A_{gb} are the cross-sections of the grains and the grain boundaries per unit area.

With conventional thickness δ of grain boundaries, $A_l \approx 1$ and $A_{gb} \approx 2\delta/d$, where d is the average grain diameter [48]. Instead of the lattice diffusion constant D_{vol}, the effective diffusivity D_{tot} must now be considered:

$$D_{tot} = D_{vol} + \frac{2D_{gb}\delta}{d} \quad (4.64)$$

Thus, the value of the diffusion "coefficient" has increased. This may also influence the regime of layer growth, in particular if the thickness of the film is small. Short-circuit diffusion may enhance the atom transport to such an extent that the reaction(s) at the interfaces become rate limiting.

Harrison [49] has defined three types of kinetics according to the ratio of grain boundary to volume diffusion, as shown in Fig. 4.23:

Type A kinetics: Strong penetration from the grain boundary to the bulk. In this case the diffusing material can escape very easily from the grain boundary into the bulk. If there are enough grain boundaries, it appears as if continuous diffusion

4.3 Microstructures

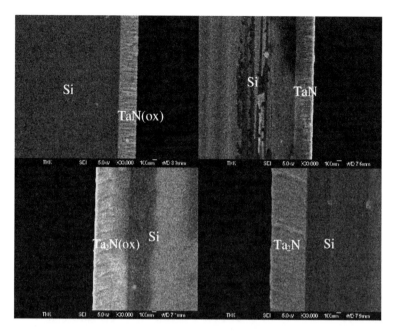

Fig. 4.22 Cross-sectional SEM-micrographs of as-deposited **a** TaN on SiO$_2$/Si, **b** TaN on silicon **c** Ta$_2$N on SiO$_2$/Si and **d** Ta$_2$N on silicon showing columnar microstructure. Micrographs are in the same scale and scale bar length is 100 nm

takes place in the grain boundary phase, corresponding to a diffusion coefficient (total), which is higher than the lattice diffusion, as shown in Eq. 4.64.

Type C: In this case there is negligible penetration from the grain boundary into the bulk. Thus, bulk and grain boundaries can be treated as two different homogeneous materials and apply Fick's laws independently.

Type B: This is a regime between A and C. Thus, there is considerable penetration from the grain boundary into the bulk. In the bulk diffusion occurs in two dimensions: the volume diffusion normal to the surface and lateral diffusion of atoms penetrating from the grain boundary into the bulk. In this case there is interdependence of grain boundary and bulk diffusion and they cannot be treated separately. Hence, new models are needed for this type of kinetics. Models for type B-kinetics have been proposed by Fisher [50], Whipple [51] and Suzuoka [52]. The latter two are modifications of the original model by Fisher.

Next we will briefly discuss the model of Fisher as it provides the simplest approach to the type B-kinetics. The model [50] takes the following view of the diffusion process: grain boundary is taken as a material with high permeability, which is then sandwiched between two large volumes of material with moderate permeability. The grain boundary (or a slab) has a thickness δ, which is so small that concentration variation across it can be taken to be negligible. As it is assumed that the volume diffusion coefficient D_v^* is much smaller than the grain boundary coefficient D_{gb}^*, the concentration in the grain boundary can be written as:

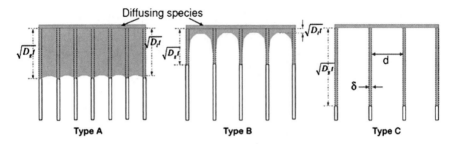

Fig. 4.23 Three different kinetic regimes according to Harrison [49]

$$\frac{\partial c_{gb}}{\partial t} = D^*_{gb}\frac{\partial^2 c_{gb}}{\partial x^2} + \frac{2}{\delta}D^*_v\frac{\partial c_v}{\partial y} \qquad (4.65)$$

where the first term describes diffusion along the grain boundary and the second term the lateral transport of matter from the grain boundary to the bulk. This treatment leads to the so-called isoconcentration curves (Fig. 4.24).

4.3.2 The Role of Impurities

Impurities have important effects on the formation of phases in thin film and bulk couples [54]. The presence of some impurity may enhance the formation of a particular phase at the expense of another. Impurities may increase or decrease reaction temperatures or influence the kinetics of a phase transformation. Impurities are also frequently responsible for the absence of phases in diffusion couples as compared to the corresponding phase diagram. One example of the increased reaction temperature is the formation of $TaSi_2$, in the reaction between thin Ta film and Si substrate, which occurs at 923 K [55]. However, if there is oxygen at the Si/Ta interface the temperature of formation will rise well above 1023 K [56]. Another example of bulk samples is the catalysing effect of phosphorous on the formation of Cu_3Si in the reaction between the bulk copper foil and the Si substrate [57]. The effect of impurities on diffusional transport should also be considered. Impurities may segregate preferably to grain boundaries and interfaces. When they segregate to grain boundaries they may reduce the effect of the short-circuit diffusion paths, thus affecting the mass transport in the system. Finally, it is emphasised that certain trace elements can have significant effects on the stability of phases in the system under investigation. Examples of this type of behaviour are given in Chap. 6.

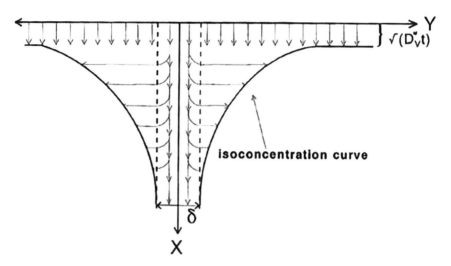

Fig. 4.24 Schematic presentation of the model of Fisher showing two-dimensional diffusion near the grain boundary [53]

4.3.3 Segregation

For a new phase to form, at least two factors are required: the thermodynamic driving force and the mobility of atoms. The latter is strongly dependent on the microstructures of the materials taking part in the reaction. As has been mentioned, thin film structures typically possess very high density of grain boundaries. These more or less disordered regions offer high diffusivity paths for the atoms. This is clearly denoted when thin film metallisations have columnar structures, which are often observed, especially after physical vapour deposition (PVD). Thus, it would be beneficial to somehow eliminate these short-circuit diffusion paths. This can be achieved, for example, by blocking the grain boundaries with impurity atoms/compounds or by producing an amorphous structure.

The driving force for the equilibrium segregation of solute or impurity atoms to grain boundaries is the systems tendency to lower its total free energy. In addition to kinetic constraints the extent of intergranular segregation depends on impurities' influence on the grain boundary energy as well as on the factors controlling their solubility, i.e. size factor and chemical interactions between dissimilar atoms. Since both the kinetics and the solubility depend on temperature, the segregation of impurities typically decreases with increasing temperature. By gathering large amounts of experimental data on grain boundary segregation Hondros and Seah [58] showed that the smaller the solubility of an impurity in the solvent the higher its segregation potential. This "rule of thumb" is frequently used when considering the segregation tendency of a given impurity.

Because of the analogies between intergranular segregation and adsorption at free surfaces, classical free surface adsorption models have often been used for evaluating grain boundary segregation [59]. This approach is valid if it takes into account the specific features, which differentiate the grain interface from surfaces. Thus, even the most dilute grain boundary cannot be regarded as a two-dimensional phase with the same components as in the bulk [60]. It is to be noted that the equilibrium condition, i.e. that the chemical potential of a component i that has the same value in all phases of the system, is valid also for grain boundaries and surfaces.

It is also possible to use Gibbs energy diagrams to investigate the segregation of a given impurity to grain boundaries. This involves a so-called constant volume condition. If we assume that the interface can be approximated as a thin layer of a homogeneous phase with constant thickness and its own Gibbs energy function as well as that the partial molar volumes of all the phases (including the interfacial phase) are independent of composition, we can use parallel tangent construction to find the interfacial composition. In this case we consider the exchange of atoms (A and B) between the interface and the bulk. The number of atoms at the interface is considered to be constant. Thus, if atom A leaves the interface and enters the bulk and atom B moves in the opposite direction at the same time the Gibbs free energy should not change:

$$\mu_A^\alpha - \mu_A^b = \mu_B^\alpha - \mu_B^b \qquad (4.66)$$

where α refers to the bulk phase and b to interface. This can be rewritten as:

$$\mu_B^b - \mu_A^b = \mu_B^\alpha - \mu_A^\alpha \qquad (4.67)$$

which gives the slopes for the interfacial phase and for the bulk phase:

$$\frac{dG_m^b}{x_B} = \frac{dG_m^\alpha}{x_B} \qquad (4.68)$$

For interphase segregation, where the volume is not necessarily fixed, other approaches like that of Gibbs surface excess model (Sect. 5.1), must be utilised. Several more advanced treatments of intergranular segregation have been published during the past decades. Extensive reviews of the models can be found from [58–60].

4.3.4 Amorphous Structures

Elimination of high diffusivity paths can also be achieved by amorphisation of the crystal structure [61–64]. It is generally thought that the absence of grain boundaries slows down the diffusion of all the elements through the barrier layer. However, since structure is not ordered, there should be more free space in amorphous materials for atoms to squeeze between others than in crystal. Thus, "volume" diffusion in amorphous materials is generally much faster than in crystalline media. Nevertheless, in thin film samples grain boundary density is usually so high that elimination of grain boundaries is anticipated to be advantageous. It is to be noted that the detailed mechanisms of diffusion in amorphous alloys are still not well understood [65].

In thin film technology there has recently been considerable advances in producing metastable and amorphous films resulting from the development of the PVD techniques, especially magnetron sputtering. During sputter-deposition, the atoms condensing in an intermixed state try to find a stable configuration [66]. Structural order in a coating is produced largely by the mobility of the adatoms. Low mobility does not allow the formation of equilibrium phases and metastable and/or amorphous phase formation is likely. Therefore phase formation and crystalline state is mainly influenced by substrate temperature together with surface and bulk diffusion [67]. It should also be noted that it is easier to form amorphous structures with mixtures/compounds than with pure elements.

Amorphous films are, however, always metastable, as pointed above. For an amorphous material of a given composition, there always exists a crystalline phase or a mixture of several crystalline phases which are thermodynamically more stable than the amorphous phase. In other words, amorphous phases exist only because nucleation or/and growth of equilibrium phases is prevented. Thus, there is generally a considerable large driving force for the crystallisation of the film. However, if the amorphous phase is fully relaxed, a discrete nucleation stage with an activation energy barrier is needed. This means that a considerable amount of energy must be provided for the system to access the lower energy state. It is also known that transition metals can be stabilised in their amorphous forms by adding metalloids (e.g. B, C, N, Si and P) [68]. This concept has been used for instance in the case of TaSiN films. The addition of Si into TaN film resulted into an amorphous structure, which did not crystallise easily [61–64].

The crystallisation temperature of the amorphous film is perhaps the most important parameter, which determines the effectiveness of the amorphous layer, for example when used as a diffusion barrier for Cu metallisation. It has been experimentally found that the presence of copper overlayer enhances the crystallisation of some underlying amorphous films [61]. As amorphous films are commonly treated as undercooled liquids, one may think that copper, which diffuses into the amorphous layer, offers heterogeneous nucleation sites for the crystallites to form, thus reducing the critical nucleus size, which is required for the formation of a stable crystal.

4.3.5 The Role of Nucleation

When a new phase AB is formed at an interface between two other phases A and B, the creation of two new interfaces A/AB and AB/B at the expense of the old A/B interface takes place. This typically leads to an increase in surface energy $\Delta\sigma$. The competition between the gain in chemical free-energy ΔG and the energy loss

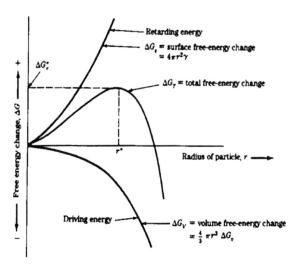

Fig. 4.25 The free energy of a nucleus as a function of its radius, showing the surface contribution (*positive*), the volume contribution (*negative*), and their sum

due to the interface term $\Delta\sigma$ is expressed as an activation energy ΔG^* for nucleation and is proportional to $\Delta\sigma^3/\Delta G^2$ (Fig. 4.25). This is a simple picture provided by the classical theory of nucleation. There has been much criticism of the classical theory of nucleation. Nevertheless, it illustrates quite simply the major factors involved in the nucleation process.

In its simplest form, classical nucleation theory starts with the equilibrium between two phases of a given substance at either the melting or evaporation point. At the equilibrium point T_c, the free energy change ΔG equals zero.

$$\Delta G = \Delta H - T_c \Delta S = 0 \tag{4.69}$$

thus leading to:

$$\Delta S = \frac{\Delta H}{T_c} \tag{4.70}$$

At the transition temperature the driving force is zero, and nothing can happen. At any temperature T_1 away from the transition temperature, the phase transformation is driven by the driving force:

$$\Delta G = \Delta S(T_1 - T_c) \tag{4.71}$$

which is valid as long as T_1 is not too far from T_c. This transformation is opposed by the surface energy contribution.

The total free energy of a nucleus with average radius r and free energy ΔG_1 (calculated per unit volume) is expressed generally as:

$$\Delta G = br^2 \sigma - ar^3 \Delta G_1 \tag{4.72}$$

where a and b are geometrical terms taking into account the fact that if the nucleus is crystalline, it will generally not be spherical because of the anisotropic

4.3 Microstructures

characteristics of crystalline elements, but tries to adapt some definite shape with minimum surface energy. The relation between the free energy of a nucleus and its radius is shown in Fig. 4.25. As seen, ΔG passes through a maximum that corresponds to the critical size r^* of the nucleus. The population of nuclei smaller than r^* will exist in some form of quasi-equilibrium distribution (e.g. they constantly appear and disappear maintaining some kind of "equilibrium" distribution), whereas nuclei bigger than r^* will tend to grow. The value of the critical nucleus can be obtained by derivation of (4.72):

$$r^* = \frac{2b\sigma}{3a} \Delta G_1 \qquad (4.73)$$

Further, the free energy of the critical nuclei becomes (Eqs. 4.71–4.73) (at any temperature T):

$$\Delta G^* = 4b^3\sigma^3 T_c^2 \Big/ 27a^2 \Delta H^2 (T - T_c)^2 \qquad (4.74)$$

At that temperature, the rate of nucleation ρ^* will be proportional to the concentration of critical nuclei and to the rate at which such nuclei can form, generally some diffusion term of the type $\exp(-Q/kT)$, so that (with a proportionality factor K):

$$\rho^* = K \exp(-\Delta G^*/kT) \exp(-Q/kT) \qquad (4.75)$$

Thus, the rate of nucleation is composed of the chemical term (driving force) and the kinetic term (activation energy for diffusion e.g. formation if critical nuclei). For example in solidification the driving force increases as the undercooling increases but at the same time the diffusion of atoms, which is required for the formation of clusters that can at some point become nuclei, slows down. Therefore, there is typically an optimum combination of the two terms, which provides the maximum nucleation rate.

When a nucleus of a new phase is formed, this generally causes a volume change to take place. In solids this is accompanied by a deformation energy loss ΔH_d (elastic energy, plus, e.g., the energy necessary to generate dislocations). The activation energy for nucleation becomes proportional to $\Delta \sigma^3/(\Delta G_c - \Delta H_d)^2$, where ΔG_c is the "chemical" free energy of bulk phases. One can derive this by first including the term for strain energy into Eq. 4.72. It is also a bit more convenient to consider the free energy per atom of the nucleus rather than the free energy per volume of the nucleus as in Eq. 4.72. The free energy associated with the formation of a nucleus of n atoms, ΔG, may be written as

$$\Delta G = n\Delta G_c + \eta n^{\frac{2}{3}}\gamma + nH_d \qquad (4.76)$$

where n is the number of atoms in nucleus, ΔG_c is the bulk (chemical) free energy change per atom in nucleus, η is the shape factor such that $\eta n^{2/3}$ equals the surface area, γ is the surface tension ($\approx$ surface free energy), and H_d is the strain energy per atom in the nucleus.

One may regroup this equation as

$$\Delta G = n(\Delta G_c + H_d) + \eta n^{\frac{2}{3}}\gamma \qquad (4.77)$$

The term ΔG_C will be negative below the transformation temperature, whereas H_d and γ are both positive. Hence if $|\Delta G_C| > H_d$ then the first term is negative. The free energy required (ΔG^*) to form the critical size nucleus (n^*) is found by differentiating Eq. 4.77 and assuming constant ΔG_C, H_d, η and γ.

$$\Delta G^* = \frac{4}{27} \frac{\eta^3 \gamma^3}{(\Delta G_c + H_d)^2} \qquad (4.78)$$

A large strain energy H_d reduces the denominator and makes ΔG^* large. This means that nucleation is more difficult since the critical size nucleus has a higher energy of formation. The lattice must compensate for this strain energy, either by atom or dislocation movement. Large strains may change the energetic status of the system and lead to formation of metastable structures (e.g. amorphous phases).

The most important parameters determining the phase selection during nucleation are the activation free energy of growth Q (or ΔG^*_2), the interface energies σ and, the chemical driving force ΔG_C and the elastic strain energy ΔH_d. In order to evaluate a possible phase selection, reasonable estimates of these parameters must be obtained.

4.3.5.1 The Activation Energy of Growth Q

This term can be approximated by the activation energy of diffusion, since the formation of a critical nucleus is mainly determined by diffusional jumps. If this is not available it can be also approximated with the activation energy of growth of the formed phase in the planar growth regime. From investigations on the later stage of growth it has been concluded that the precipitates formed in the A/B interface first grow to coalescence within the A/B interface [69]. These results indicate that growth of the nucleus is preferred in the direction of the interface emphasizing the importance of the atomic mobility of the A/B interface. Therefore, it is reasonable to assume that the activation energy of volume diffusion is only an upper estimate for the activation energy of growth.

4.3.5.2 The Interface Free Energy σ

The interface free energies of crystalline phases consist of two contributions: the chemical contribution related to (chemical) atomic interaction energy, and the structural contribution which originates from the free energy of structural defects associated with semicoherent and incoherent interfaces [70]. The interface free energy terms are hardly known, and even if they are known they are usually bulk values and therefore, their use in early stages of phase formation is highly questionable. The interfacial free energies in the solid state are likely to vary between values approaching zero for epitaxial interfaces to maximum values of the order of 2000 ergs/cm^2 for random interfaces [71]. It should be emphasised that surface free energies of crystalline solids are very much dependent on the history of the specimen, crystal orientation, defect density, impurities etc. Hence, the values of surface free energies for solids are not material constants but are highly specific for the sample in question.

4.3.5.3 The Elastic Strain Energy ΔH_d

Elastic strain energy affects the nucleation as can be seen from Eq. 4.75. Here, one will consider only elastic energy, yet the deformations involved can reach proportions beyond the usual elastic limits of the materials. A complete analysis of these effects should take into account the energy stored in various defects owing to plastic deformation as well. The elastic energy resulting from the formation of a third phase at the two-phase interface will depend on the elastic characteristics of all three phases. In the simplified case where all three phases have the same elastic constants, the energy is given by the following relation [72, 73]

$$\Delta g_{el} = [2s(1+v)/9(1+v)]\varepsilon^2 \qquad (4.79)$$

where s, the shear modulus of elasticity, is also called the modulus of rigidity, v is the Poisson ratio, and ε, the strain, is the ratio between the excess volume (under zero stress) and the volume of the hole (here the volume of the reactants) [71]. Ignoring the problems of anisotropy, the following relations from the theory of elasticity can be utilised. Young's modulus E is equal to s times $2(1+v)$; some tables also give the compression or bulk modulus, which is equal to E, divided by $3(1-2v)$. Typically, the problems of interest do not concern isotropic materials, but three different materials at once. Nevertheless, as a first approximation one may consider that the strain energy is given by an average of the different elastic constants.

4.3.5.4 The Chemical Driving Force

In order to calculate the chemical driving force for the nucleation of a new phase at the A/B interface, the free energy curves of the solid solution phases and of the formed compounds must be known. These can be determined by applying the CALPHAD method [19–21] as already discussed. In systems where the compounds that are forming are stoichiometric with large negative heats of formation, the driving force for the nucleation of a new phase from the pure elements is large and equal to the free energy of compound formation at the particular reaction temperature. However, in most systems interdiffusion occurs first or

Fig. 4.26 Schematic illustration of the free energy versus composition of a system A–B, where elements A and B have different equilibrium crystal structures α and β, respectively, and they can form a crystalline intermetallic phase η and a metastable intermetallic phase M [74]

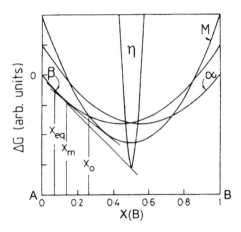

simultaneously to compound formation, thus substantially decreasing the initially available driving force for compound formation [74].

In Thompsons [74] treatment it is suggested that interdiffusion not only might precede, but also must precede nucleation of new phases, and that only after some interdiffusion has occurred can there be a driving force for nucleation. This requirement imposes a kinetic constrain on the first phase formation. Thus, the relative mobilities of diffusing components in competing phases will determine which phase will form first.

In Fig. 4.26 the Gibbs energy curves of the A–B system at a given temperature are shown, where A and B have different crystal structures α and β, respectively, and they can form a stable intermediate phase η and a metastable phase M. By assuming that B diffuses into β much faster than A into α, we simplify the case so that only diffusion of B into β needs to be considered. According to the thermodynamic principle of phase equilibrium, as diffusion proceeds, the first phase to nucleate is η in β when a sufficient volume of β has reached its equilibrium composition X_{eq} with η (see the tangent line between β and η in Fig. 4.26). If η cannot nucleate at this moment, it will become possible for the metastable phase M to nucleate in β when a sufficient volume of β has reached its equilibrium composition X_m with M, provided that the interdiffusion continues. The appearance of the metastable phase M would therefore indicate that the time required for nucleation of η is longer than the time required for interdiffusion to the point at which M can nucleate. If neither η nor M can nucleate when the composition of β passes through X_{eq} and X_m sequentially, β will polymorphically transform into M when its composition reaches X_0. This analysis leads to a conclusion that the phase selection depends on the interdiffusion rate in the parent phases as well as the nucleation rates of the product phases. A similar case was treated in Sect. 4.1.8 where the dissolution of Cu into liquid Sn took place and the solid state diffusion of Sn into Cu could be neglected. This type of analysis is used extensively together with ternary phase diagrams in Chap. 6.

4.3 Microstructures

While the relative rates of nucleation are ultimately controlling the phase selection, which phase can nucleate (and grow) is controlled by interdiffusion [75]. Hence, the first phase formation depends on the kinetics. And the nucleation rates are controlled not only by the barriers to nucleation, and hence the volume Gibbs energy ΔG_v and the energies of the interfaces σ involved, but also by the diffusion required to form critically sized clusters of the product phases. Thompson [74] further suggested that if one component diffused rapidly into the other and self-diffusion of the host was slow, polymorphic phase transitions were favoured, so that phases with broad compositional ranges of stability and phases that were rich in the slowly moving component were favoured, and such phases with low energy interfaces with the host phase(s) would be further favored.

It should also be emphasised that the free energies of compound formation calculated by the CALPHAD method refer to bulk materials. In a small nucleus the chemical long-range order may not be fully developed due to the interfacial constraints thus increasing its free energy with respect to the bulk value. Therefore, the free energy of an ordered chemical compound (as derived from the CALPHAD method) should be viewed as the lower limit for the free energy of the compound nucleus. Thus, the driving force for the formation of a certain compound may be considerably smaller than calculated by using bulk values.

4.3.6 The Role of Steep Concentration Gradients

In thin film couples there often occur very steep concentration gradients. Desré and Yavari attributed the formation of the amorphous phase in a thin film system to a large composition gradient [76, 77]. By using simple thermodynamic arguments, they showed that sharp composition gradients increased the stability of an amorphous phase layer by eliminating or reducing the driving force $\Delta G_{a \to c}$ for nucleation of crystalline intermetallic phases in an amorphous layer. This effect increases with increasingly negative ΔH_{mix} and free energy of alloying ΔG_{mix}. As diffusive mixing proceeds further during the growth of the amorphous layer, the composition gradient flattens out and the driving force for crystalline phase formation is gradually restored. The energy barrier for their nucleation also diminishes towards the value of the classic theory.

The existence of a composition gradient in the interfacial region has been experimentally observed in the Ni–Si system and the width of the interfacial region was estimated to be $\leq 2 \times 10^{-3}$ μm [78]. This region can be an amorphous phase or a crystalline solution phase. Figure 4.27b shows the Gibbs energy curve of such an amorphous region (G_{am}) at the A/B interface.

According to the classic theory of homogeneous nucleation, in order for a crystalline stoichiometric phase $\beta - A_{C^*}B_{1-C^*}$ to nucleate and grow inside the amorphous region, the nucleus with radius r in Fig. 4.27 a must attain a critical radius r^*. At this point, local equilibrium prevails at the interface between the amorphous phase and β. Let us consider first the case when the composition

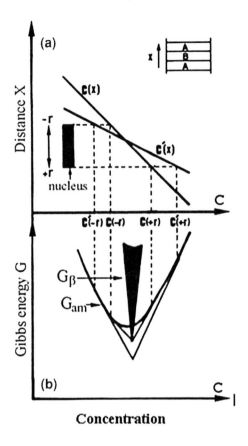

Fig. 4.27 Gibbs energy tangent constructions for compositions at the tips of the critical nucleus of intermetallic phase in an amorphous layer subject to the concentration gradient ∇C [76] (© 1991 American Physical Society)

gradient in the amorphous phase is such as represented by line C(X) in Fig. 4.27a. One can see from Fig. 4.27a that the common tangent for G_{am} and the β phase, G_β, can be drawn for both compositions C(−r) and C(+r) of the layered structure. Thus, this means that the radius r of the crystalline β-phase has reached the critical size (r = r*), local equilibrium is established and the β-phase can start to grow. However, if the composition gradient in the amorphous phase is steeper as represented by C'(X), the nucleus will not attain the critical size because the tangents both from C'(−r) and C'(+r) miss the tip of the G_β curve and the local equilibrium is not established. Thus, there exists a critical composition gradient ∇C_c for β to be able to nucleate. As long as the composition gradient in the amorphous phase is greater than ∇C_c, the crystalline phase will not form.

Following the thermodynamic approach of Cahn and Hilliard for a non-uniform system [79], the Gibbs free energy of a volume v of an amorphous layer can be written as

$$G_a(v) = \rho \int_v \left[G_0(C) + N_A \chi (\nabla C)^2 \right] dv \quad (4.80)$$

where ρ is the number of moles of atoms per unit volume, $G_0(C)$ is the Gibbs free energy per atom of an amorphous phase with uniform composition C, N_A is Avogadro's number, and χ is a constant. The first term is the "bulk" free energy and the other is the gradient term. From this the Gibbs free energy of formation of a nucleus of the compound $A_{C*}B_{1-C*}$ which includes the classic interfacial term can be calculated:

$$\Delta G_N = 24\sigma_{pc} r^2 + 8\rho \left[\Delta G_{pc}(C^*) - N_A \chi (\nabla C)^2 \right] r^3 + \tfrac{4}{3} \rho \alpha (\nabla C)^2 r^5 \quad (4.81)$$

where σ_{pc} is the interfacial energy at the polymorphous crystallisation front, $\Delta G_{pc}(C^*)$ is the Gibbs free energy of polymorphous crystallization at C^*, $\alpha = \partial^2 \Delta G / \partial C^2$ is assumed to be constant in the range $C^* - r\nabla C < C < C^* + r\nabla C$.

As can be seen from Eq. 4.81 the concentration gradient leads to a term to the fifth power of the embryo size. As the coefficient of the r^5 is positive this term will contribute to an increase in ΔG_N.

For the typical composition gradients which occur during the solid-state amorphous reaction ($\nabla C \leq 10^{+6}$/cm) the energy factor $N_A \chi (\nabla C)^2$ in Eq. 4.81 is negligible. After some algebra, an analytical expression for the critical gradient ∇C_c, at which ΔG_N is both minimum and equal to zero, is obtained as

$$\nabla C_c = \frac{\rho}{9\sigma_{pc}} \frac{(2|\Delta G_{pc}|)^{3/2}}{\alpha^{1/2}} \quad (4.82)$$

Above this critical gradient there is no driving force for compound nucleation.

4.4 Summary

The basic concepts involved in the thermodynamic-kinetic method have been introduced in this chapter. Several additional factors, such as microstructure and surfaces, which have to be taken into account to extend the analysis, have also been briefly discussed. The nature of the present method is completely general and can be used to tackle a wide range of different problems if the pertinent data is available. The concepts introduced in this chapter will be extensively utilised in the case studies presented in Chap. 6. Thus, while going through the given examples it might be worthwhile to sometimes return to the concepts discussed here.

References

1. J. Kivilahti, The chemical modeling of electronic materials and interconnections. JOM J. Miner. Metal. Mater. Soc. **54**, 52 (2002)
2. J.W. Gibbs, *The scientific papers*, vol. 1, Thermodynamics (Dover, New York, 1961)
3. L. Darken, R. Gurry, *Physical Chemistry of Metals* (McGraw-Hill, New York, 1953)

4. E.A. Guggenheim, *Thermodynamics* (Elsevier Science, The Neatherlands, 1967)
5. G.N. Lewis, M. Randall, *Thermodynamics*, rev. by K. Pitzer, L. Brewer (McGraw-Hill, New York, 1961)
6. M. Hillert, The Uses of Gibbs Free Energy-Composition Diagrams, in *Lectures on the Theory of Phase Transformations*, ed. by H.I. Aaronson (The Metallurgical Society of the AIME, Warrendale, 1975)
7. N.A. Gokcen, *Thermodynamics* (Techscience inc, Hawthorne, 1975)
8. R. Defay, I. Prigogine, *Surface Tension and Adsorption*, trans. by D. Everett (Longmans, London, 1966)
9. D. Turnbull, Metastable structures in metallurgy. Metall. Trans. A **12**, 695 (1981)
10. A.L. Greer, Transformations of metastable phases. Phil. Mag. B **61**, 525 (1990)
11. A. Prince, *Alloy Phase Equilibria* (Elsevier, New York, 1966)
12. D. Gaskell, *Introduction to Metallurgical Thermodynamics* (McGraw-Hill, Tokyo, 1973)
13. E.A. Guggenheim, *Mixtures* (Oxford University Press, London, 1952)
14. T. Massalski, *Binary Alloy Phase Diagrams* (ASM, Materials Park, 1996)
15. F. Rhines, *Phase Diagrams in Metallurgy: Their Development and Applications* (McGraw-Hill, New York, 1956)
16. A.D. Pelton, H. Schmalzried, Phase diagrams. Progr. Solid. St. Chem. **10**, 119 (1976)
17. T. Laurila, K. Zeng, J. Molarius, I. Suni, J.K. Kivilahti, Chemical stability of Ta diffusion barrier between Cu and Si. Thin Solid Films **373**, 64 (2000)
18. T. Laurila, K. Zeng, J. Molarius, I. Suni, J.K. Kivilahti, Failure mechanism of Ta diffusion barrier between Cu and Si. J. Appl. Phys. **88**, 3377 (2000)
19. T. Laurila, K. Zeng, J. Molarius, I. Suni, J.K. Kivilahti, Effect of oxygen on the reactions in the Si/Ta/Cu metallization system. J. Mater. Res. **16**, 2939 (2001)
20. L. Kaufman, H. Bernstein, *Computer Calculation of Phase Diagrams* (Academic Press, New York, 1970)
21. M. Hillert, *Phase Equilibria, Phase Diagrams and Phase Transformations : Their Thermodynamic Basis* (Cambridge University Press, Cambridge, 1998)
22. H. Lukas, S. Fries, B. Sundman, *Computational Thermodynamics—The Calphad Method* (Cambridge University Press, Cambridge, 2007)
23. I. Ansara, Thermodynamic modelling of solution and ordered phases. Pure Appl. Chem. **62**, 71 (1990)
24. K. Hack, *The SGTE Casebook—Thermodynamics at Work* (CRC Press, Boca Raton, 2008)
25. Z.-P. Jin, C. Qiu, An experimental study and thermodynamic evaluation of the Fe-Mo-Ti system at 1000°C. Metall. Trans. A **24**, 2137 (1993)
26. W. Mayr, W. Lengauer, P. Ettmayer, D. Rafaja, J. Bauer, M. Bohn, Phase equilibria and multiphase reaction diffusion in the Cr-C and Cr-N systems. J. Phase Equil. **20**, 35 (1999)
27. G. Ottaviani, Review of binary alloy formation by thin film interactions. J. Vac. Sci. Technol. **16**, 1112 (1979)
28. K.N. Tu, Interdiffusion and reaction in bimetallic Cu–Sn thin films. Acta Metall. **21**, 347 (1973)
29. P.W. Dehaven, G.A. Walker, N.A. O'Neil (eds.) in *Proceedings of Adances in X-ray Analysis*, vol. 27, 1–5 August 1983, Snowmass, Colo., USA (Plenum Publication, New York, 1984), p. 389
30. J.O.G. Parents, D.D.L. Chung, I.M. Bernstein, Effects of intermetallic formation at the interface between copper and lead-tin solder. J. Mater. Sci. **23**, 2564 (1988)
31. K.N. Tu, R.D. Thompson, Kinetics of interfacial reaction in bimetallic Cu–Sn thin films. Acta Metall. **30**, 947 (1982)
32. K. Zeng, R. Schmid-Fetzer, in *Proceedings of Werkstoffwoche '96*, 28–31 May 1996, Stuttgart, Germany, ed. by F. Aldinger, H. Mughrabi (DGM, Frankfurt, 1997), p. 75
33. R. Beyers, R. Sinclair, M.E. Thomas, Phase equilibria in thin-film metallizations. J. Vac. Sci. Technol. **B 2**, 781 (1984)

34. C. Michaelsen, S. Wöhlert, R. Bormann (eds.), in *Proceedings of Polycrystalline Thin Films: Structure, Texture, Properties and Applications*, vol. 343, 4–8 April 1994, San Francisco, California, USA (Materials Research Society, Pittsburgh, 1994), p. 205
35. K. Rönkä, F.J.J. van Loo, J.K. Kivilahti, A diffusion-kinetic model for predicting solder/conductor interactions in high density interconnections. Metal. Mater. Trans. A **29A**, 2951 (1998)
36. P. Shewmon, *Diffusion in Solids*, 2nd edn. (TMS, Warrendale, 1989)
37. J. Philibert, *Atom movements Diffusion and mass transport in solids* (Les Editions de Physique, Les Ulis, 1991)
38. C. Wagner, The evaluation of data obtained with diffusion couples of binary single-phase and multiphase systems. Acta Metall. **17**, 99 (1969)
39. K. Denbigh, *The Principles of chemical equilibrium*, 3rd edn. (Cambridge University Press, Cambridge, 1978)
40. F.J.J. van Loo, Multiphase diffusion in binary and ternary solid-state systems. Prog. Solid St. Chem. **20**, 47 (1990)
41. L.E. Trimble, D. Finn, A. Cosgarea Jr., A mathematical analysis of diffusion coefficients in binary systems. Acta Met. **13**, 501 (1965)
42. P.L. Gruzin, Self-diffusion in gamma iron. Dokl. Akad. Nauk S.S.S.R. **86**, 289 (1952)
43. A. Paul, A.A. Kodentsov, F.J.J. van Loo, On diffusion in the [beta]-NiAl phase. J. Alloys Comp. **403**, 147 (2005)
44. J.S. Kirkaldy, Diffusion in multicomponent metallic systems I. Can. J. Phys. **36**, 899 (1958)
45. J.S. Kirkaldy, L.C. Brown, Diffusion behaviour in ternary, multiphase systems. Can. Met. Q. **2**, 89 (1963)
46. D. Coates, J.S. Kirkaldy, *Proceedings of Second International Conference on Crystal Growth* (North-Holland Publishing, Amsterdam, 1968), pp. 549–554
47. K. Bhanumurthy, R. Schmid-Fetzer, Phase stability and related interfacial reactions in the Cr–Si–C system. Z. Metallkd. **87**, 61 (1996)
48. J. Philibert, Reactive diffusion in thin films. Appl. Surf. Sci. **53**, 74 (1991)
49. L.-G. Harrison, Influence of dislocations on diffusion kinetics in solids with particular reference to the alkali halides. Trans. Faraday Soc. **57**, 1191 (1961)
50. J.C. Fisher, Calculation of diffusion penetration curves for surface and grain boundary diffusion. J. Appl. Phys. **22**, 74 (1951)
51. R.T.P. Whipple, Concentration contours in grain boundary diffusion. Phil. Mag. **45**, 1225 (1954)
52. T. Suzuoka, Lattice and grain boundary diffusion in polycrystals. Trans. Jpn. Inst. Met. **2**, 25 (1961)
53. J. Gulpen, Reactive phase formation in the Ni-Si system, Doctoral Thesis, Technical University of Eindhoven, The Netherlands (1995)
54. T. Laurila, V. Vuorinen, M. Paulasto-Kröckel, Impurity and alloying effects on interfacial reaction layers in Pb-free soldering. Mater. Sci. Engin. R **R68**, 1–38 (2010)
55. G. Ottaviani, Metallurgical aspects of the formation of silicides. Thin Solid Films **140**, 3 (1986)
56. A. Christou, H.M. Day, Silicide formation and interdiffusion effects in Si–Ta, SiO_2–Ta and Si–PtSi–Ta thin film structures. J. Electr. Mater. **5**, 1 (1976)
57. J.G.M. Becht, The Influence of Phosphorous on the Solid State Reaction Between Copper and Silicon or Germanium, Doctoral Thesis, Technical University of Eindhoven, The Netherlands (1987)
58. E. Hondros, M. Seah, Segregation to interfaces. Int. Met. Rev. **22**, 262 (1977)
59. E. Hondros, M. Seah, The theory of grain boundary segregation in terms of surface adsorption analogues. Metall. Trans. A **8A**, 1363 (1977)
60. M. Guttmann, Grain boundary segregation, two dimensional compound formation, and precipitation. Metall. Trans. A **8A**, 1383 (1977)
61. P. Pokela, Amorphous Diffusion Barriers for Electronic Device Applications, Doctoral Thesis, Helsinki University of Technology (1991)

62. E. Kolawa, P. Pokela, J. Reid, J. Chen, R. Ruiz, M.-A. Nicolet, Sputtered Ta-Si-N diffusion barriers in Cu metallizations for Si. IEEE Electr. Dev. Lett. **12**, 321 (1991)
63. E. Kolawa, J.M. Molarius, C.W. Nieh, M.-A. Nicolet, Amorphous Ta–Si–N thin-film alloys as diffusion barrier in Al/Si metallizations. J. Vac. Sci. Tech. **A8**, 3006 (1990)
64. J.M. Molarius, E. Kolawa, K. Morishita, M.-A. Nicolet, J.L. Tandon, J.A. Leavitt, L.C. McIntyre Jr., Tantalum-based encapsulants for thermal annealing of GaAs. J. Electrochem. Soc. **138**, 834 (1991)
65. M. Kijek, M. Ahmadzadeh, B. Cantor, R.W. Cahn, Diffusion in amorphous alloys. Scr. Metall. **14**, 1337 (1980)
66. H. Holleck, M. Lahres, P. Woll, Multilayer coatings—influence of fabrication parameters on constitution and properties 1. Surf. Coat. Tech. **41**, 179 (1990)
67. O. Knotek, A. Barimani, F. Löffler, in *Thin film structures and phase stability : Symposium*, 16–17 April 1990, San Francisco, California, USA, ed. by B. M. Clemens
68. M.A. Nicolet, in *Diffusion in Amorphous Materials,* ed. by H. Jain, D. Gupta (TMS, Warrendale, 1994), pp. 225–234
69. K.R. Coffey, L.A. Clevenger, K. Barmak, D.A. Rudman, C.V. Thompson, Investigating the thermodynamics and kinetics of thin film reactions by differential scanning calorimetry. Appl. Phys. Lett. **55**, 852 (1989)
70. R. Bormann, Kinetics of interface reactions in polycrystalline thin films. Mat. Res. Soc. Symp. Proc. **343**, 169 (1994)
71. F.M. d'Heurle, Nucléation of a new phase from the interaction of two adjacent phases: Some silicides. J. Mater. Res. **3**, 167 (1988)
72. K.C. Russel, Nucleation in solids: the induction and steady state effects. Colloid Interface Sci. **13**, 205 (1980)
73. J.D. Eshelby, The determination of the elastic field of an ellipsoidal inclusion, and related problems. Proc. R. Soc. Lond. Ser. A **241**, 376 (1957)
74. C.V. Thompson, On the role of diffusion in phase selection during reactions at interfaces. J. Mater. Res. **7**, 367 (1992)
75. T. Laurila, J. Molarius, Reactive phase formation in thin film metal/metal and metal/silicon diffusion couples. Crit.ll Rev. Solid State Mater. Sci. **28**, 185 (2003)
76. P.J. Desre, A.R. Yavari, Suppression of crystal nucleation in amorphous layers with sharp concentration gradients. Phys. Rev. Lett. **64**, 1533 (1990)
77. P.J. Desre, Effect of sharp concentration gradients on the stability of a two-component amorphous layer obtained by solid state reaction. Acta Metall. Mater. **39**, 2309 (1991)
78. P.J. Grunthaner, F.J. Grunthaner, J.W. Mayer, XPS study of the chemical structure of the nickel/silicon interface. J. Vac. Sci. Technol. **17**, 924 (1980)
79. W. Cahn, J.E. Hilliard, Free energy of a nonuniform system. I. Interfacial free energy. J. Chem. Phys. **28**, 258 (1958)

Chapter 5
Interfacial Adhesion in Polymer Systems

Delamination is the manifestation of poor interfacial adhesion between dissimilar materials. Electronics components are manufactured using various dissimilar materials and these components are assembled into functional devices using again various dissimilar materials. Inherently many of the employed materials repel each other's proposing adhesion problems. Second, dissimilar materials respond differently to environmental stress factors like temperature and humidity changes, mechanical shocks etc. The above-mentioned facts result in a situation that demands in-depth understanding of interfacial adhesion when reliably performing electronics, microelectronics and bioMEMSs are designed.

There is no doubt that molecules or atoms of solid materials stick together. Also the mechanisms by which this attraction between materials takes place have been described thoroughly in the literature. Nevertheless, there is a fundamental difference between adhesion as a molecular level phenomenon (adsorption, electrical attraction) and measured adhesion as an engineering level experience. At the engineering level, other adhesion mechanisms (being mechanical interlocking and diffusion) contribute significantly to the practical endurance and/or strength of the interface. These mechanisms often contribute simultaneously and result in some measurable adhesion that can be explained to a certain extent mathematically for simple interfaces utilising concepts like the thermodynamic work of adhesion and continuum fracture mechanics as will be shown later. In practice, the situation is far more complicated and it will be hard, if not impossible, to find a relation that would also consider starting with the following things: mechanical aspects (properties of the joint materials, geometry of the interface and roughness of the interface), defects at the interface (air bubbles, reaction products, degradation products, weak boundary layers, etc.), and penetration of foreign molecules into the interface (water through the interface or bulk material). The following two (apparently) simple questions are not easy to answer: first, *will any two given materials stick together?*; and, second, if they do stick together, *how strong will they remain stuck together over the lifetime of the product?*

Interfacial adhesion may be described as the force needed to separate two bodies along their interface, and it is restricted therefore to the interfacial forces acting across the interface. A value for interfacial adhesion is commonly reported if the fracture path results in delamination and hence does not penetrate deeply into the bulk of one of the joint materials. When it does, the failure mechanism is said to be cohesive. Since the failure mechanism and adhesion strength never depend solely on the forces acting across the interface, the term "practical adhesion" is sometimes favored. In the following text the term interfacial adhesion is used except where there is a clear indication of cohesive failure.

In order to understand interfacial adhesion better we need to consider several aspects of surface and interfacial phenomena, such as adsorption, surface free energy, the kinetics of wetting and examine some processing, testing, inspection and environmental stressing methods. This chapter discusses these concepts and we try to give some theoretical means to better understand, test and design reliable interfaces using dissimilar materials.

5.1 Adsorption

Adsorption of gaseous species to metal surfaces was mentioned already in the previous section. Owing to its importance for small scale structures, the phenomenon will be briefly discussed in the present chapter. A solid surface in contact with a gaseous atmosphere is, in general, instantly covered with an adsorbed layer of gas molecules. This process is, of course, accompanied by the overall decrease in the Gibbs energy of the system. The adsorption of a gas on solid surface may involve only physical interactions (due to van der Waals forces), or there may be a chemical interaction with the formation of chemical bonds between the solid and the gas. The former is called physical adsorption and the latter chemical adsorption. The distinction between the two forms of adsorption is important from the fundamental and application point of view.

Physical adsorption involves only secondary forces between the solid and the gaseous species, thus making it similar to the condensation of a liquid from the vapour. The enthalpy change related to the physical adsorption is therefore of similar magnitude to that of the liquefaction. On the contrary, in chemical adsorption primary bonds are formed during the process and thus the enthalpies involved are much higher. Next some general differences between physical and chemical adsorption are discussed in brief [1].

5.1.1 Enthalpy of Adsorption

The enthalpy of physical adsorption is typical if the same order as that of condensation (well below 20 kJ/mol), whereas that of chemical adsorption is much higher (rarely below 80 kJ/mol). This difference is easily rationalised by the fact that in physical adsorption interactions are secondary and in chemical adsorption primary.

5.1.2 Specificity

Since chemical adsorption is accomplished by chemical reactions between gas molecules and surface groups on a solid, it is highly specific and thus depends on the combination of the solid and the gas. On the contrary, physical adsorption is nonspecific and takes place with any gas–solid combination under the appropriate temperature and pressure conditions.

5.1.3 Reversibility

As only secondary interactions are involved in physical adsorption, it can be reversed (gas desorbed) by lowering the gas pressure and usually without raising the temperature. The kinetics of the process can, however, be very slow. The desorption of chemically adsorbed gas requires the breaking of chemical bonds making it much more difficult. Low pressures combined with elevated temperatures are typically required.

5.1.4 Thickness of the Adsorbed Layer

Chemisorption is typically limited to one layer due to its site specificity, whereas in physical adsorption (due to its non-specific nature) several layers are in general adsorbed. It is also possible that physical adsorption takes place on the top of the chemically adsorbed layer, which may obscure the nature of the adsorption of this first layer.

5.1.5 Conditions of Adsorption

The experimental conditions that favour chemical adsorption are typically high temperatures and a wide range of temperatures in contrast to low temperatures and pressures close to those of condensation, which favour physical adsorption.

Adsorption can be addressed in thermodynamic terms by utilising the Gibbs surface excess approach, which takes when T is constant the form:

$$Ad\gamma = -\sum_i n_i^s d\mu_i^\phi \tag{5.1}$$

$$d\gamma = -\sum_i \Gamma_i^s d\mu_i^\phi \tag{5.2}$$

where $\Gamma_i^s \equiv \frac{n_i^s}{A}$ is the excess surface concentration.

In equilibrium $\mu_i^\phi = \mu_i^s = \mu_i$ so

$$d\gamma = -\sum_i \Gamma_i^s d\mu_i \qquad (5.3)$$

This equation gives the relation between the surface excess of a species i and the surface energy of the given phase. Using this equation it is easy to understand that when there are species in the system that can lower the surface energy or tension they will tend to accumulate on the surface (are surface active). This will result in an overall decrease in the systems total Gibbs free energy by decreasing the surface energy contribution. It is to be noted that the Gibbs surface is determined as a sharp mathematical boundary between two bulk phases α and β. It is also assumed that the properties of the bulk phases are retained right up to the boundary. Thus, it is obvious that the location of the dividing surface will have an influence on the magnitude of the surface excess concentration. There are several conventions to determine the location of the dividing surface. One of them was suggested by Gibbs himself and its called the Gibbs convention. In this convention the surface is placed so that the surface excess of the major component is zero. An approach which tries to avoid the somewhat artificial concept of the Gibbs dividing surface has been developed by Guggenheim [2]. In this concept the surface region is supposed to be a bulk region whose upper and lower limits lie somewhere in the α and β phases and not too far from the actual surface.

Finally, two equations, the Laplace and the Kelvin equations, are briefly introduced here as they have a central role when small scale structures and surfaces are considered. If a fluid interface is curved between two phases then it turns out that the pressures on either side must be different. When the system is in equilibrium, every part of the surface must be in mechanical equilibrium (also in thermal and chemical). For a curved surface the forces of surface tension are exactly balanced by the difference in pressure on the two sides of the interface. This is expressed by the Laplace equation:

$$P^\alpha - P^\beta = \gamma \left(\frac{1}{r'} + \frac{1}{r''} \right) \qquad (5.4)$$

where P stands for the pressure, γ is the surface tension and r' and r'' are the radii of curvatures. By convention, positive values are assigned for the radii of curvature if they lie in phase α (see Fig. 5.1).

An important consequence of the Laplace equation concerns the effect of surface curvature on the vapour pressure of a liquid. This relationship is known as the Kelvin equation:

$$\ln\left(\frac{p^c}{p^\infty}\right) = \left(\frac{\gamma V_m^L}{RT}\right)\left(\frac{2}{r_m}\right) \qquad (5.5)$$

5.1 Adsorption

Fig. 5.1 Forces on a spherical cap

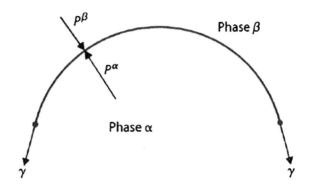

where p^c and p^∞ are the vapour pressures over the curved surface of mean curvature r_m ($\frac{1}{r_m} = \frac{1}{2}(\frac{1}{r'} + \frac{1}{r''})$) and a flat surface ($r = \infty$) and V_m^L is the molar volume of the liquid. There is a convention to assign a positive sign to r_m when it lies in the liquid phase and a negative sign when it lies in the vapour phase. Equation 5.5 can be used, for example, to rationalise capillary condensation. Condensation occurs when the actual vapour pressure exceeds the equilibrium vapour pressure. If the surface is curved, the Kelvin equation 5.5 shows that the actual pressure can be significantly lower than equilibrium pressure and thus condensation in pores in the solid (if the liquid wets the solid) or between closed spaced solid particles may occur.

5.2 Surface Energy and the Wetting of a Solid Polymer Surface

In most cases, intimate molecular contact at an interface is a necessary requirement for the development of strong interfacial adhesion. This means that the material needs to spread onto the substrate and wet it. Wetting will take place when the surface free energy of the solid substrate is greater than the sum of the surface tension of the liquid and the interfacial free energy. The surface free energy of polymers is comparatively low in relation to that of many other materials and problems often arise in the wetting of solid polymer surfaces. In the case of physical retention by mechanical interlocking, sufficient adhesion may be provided without good wettability, but such an interface is highly susceptible to destructive environmental processes resulting in interfacial failures.

In understanding surface free energy, it is useful to consider a liquid drop resting on a solid surface, as depicted in Fig. 5.2. In equilibrium and in the absence of any external forces, the drop will spontaneously assume the form of a sphere. This shape corresponds to the minimum surface-to-volume ratio. It can be assumed that work must be done on the drop to reshape it and therefore increase its surface-to-volume ratio. Also, the molecules in the surface are assumed to be at

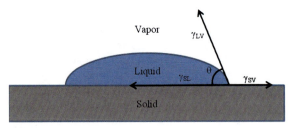

Fig. 5.2 A liquid drop on a solid surface and the surface energy balance. The γ_{lv} is surface free energy of the liquid, γ_{sl} is the solid/liquid interfacial surface free energy and γ_{sv} is the surface free energy of the solid. θ is the contact angle

higher energy state than those in the bulk liquid. This can be explained by the fact that a surface molecule has fewer neighbours and, as a consequence, less intermolecular interaction than bulk molecules. There is, then a free energy change associated with the isothermal reversible formation of a liquid surface and this is termed the surface free energy or surface tension. It must be emphasised that this surface free energy is not the total free energy of the surface molecules but rather the excess free energy that the molecules possess by virtue of their being in the surface.

The surface free energy of liquids can be measured directly because the surface formation is reversible and the molecules attain equilibrium almost as soon as a new surface is formed. In principle, all that has been said above for liquids also applies to solid surfaces. In practice, the situation is rather different. This is because in solids, unlike liquids, the surface tension and surface energy are not equal. In solids surface energy has an extra component, the so-called surface stress. In liquids the surface area is increased by transporting atoms from the bulk to the surface. On the other hand, in solids the surface area is increased by stretching the "old" surface without the addition of extra atoms. This results in surface stress and the fact that surface tension and surface energy are not equal for solids. In the literature, various approaches are presented that enable the evaluation of the solid surface free energy with measured contact angles by liquids with known surface tension [3–22]. All the models derived from these approaches are based on Young's equation shown below, and according to the model used, the measurement of one, two, three or more probing liquid contact angles on the solid are needed:

$$\gamma_{lv} \cos \theta = \gamma_{sv} - \gamma_{sl} \qquad (5.6)$$

In this case, the known parameters in Young's equation are γ_{lv} and θ, and it is expected that the solid–vapour surface free energy (γ_{sv}) can be provided by this equation. However, the solid–liquid surface free energy (γ_{sl}) is unknown. If this parameter could be expressed by another equation in terms of γ_{lv} and γ_{sv}, the problem could be solved.

Three basic approaches have been taken to solve this problem. The first approach aims to find a mathematical relationship for γ_{sl} in terms of γ_{lv} and γ_{sv} and then evaluate the unknown γ_{sv}. This approach has led to the formulation of several

5.2 Surface Energy and the Wetting of a Solid Polymer Surface

equation-of-state models, of which the latest modification has been presented very recently [13]. The second approach approximates the surface free energy of a solid with the concept of a critical contact angle [10]. The models based on this approach assume that the surface tension of a liquid that exhibits a zero contact angle on the solid surface equals γ_{sv}. However, these models generally yield values lower than those of the models based on the other approaches and require the use of many homologous probing liquids for the determination of γ_{sv}, which is time-consuming. Because of these drawbacks, the second group of models is not considered further in this chapter. The approaches belonging to the third group were originally developed to provide an explanation of the results obtained within the context of the critical surface angle described previously [10]. These approaches divide the surface free energy into different components [14, 15]. The value γ_{sl} is then expressed in terms of the components of γ_{lv} and γ_{sv}. Models formulated according to this approach include the geometric mean [16, 17] harmonic mean [20], and acid–base [21–32] models, and the determination of γ_{sv} with these models requires the measurement of contact angles of two to three liquids on a given solid with known components of the surface tension. Here most of the attention will be given for the acid–base model, as it provides the most complete picture of the factors contributing to the surface energy of solids.

5.2.1 Equation-of-State Model

If the values of γ_{lv}, γ_{sv}, and γ_{sl} are constant during an experiment and $\gamma_{lv} \cos \theta$ depends only on γ_{lv} and γ_{sv}, a mathematical relationship from such experimental contact-angle patterns should be deducible:

$$\gamma_{lv} \cos \theta = f(\gamma_{lv} \gamma_{sv}) \tag{5.7}$$

The combination of Eq. 5.7 with Young's equation 5.6 leads to the equation-of-state relationship:

$$\gamma_{sl} = f(\gamma_{lv} \gamma_{sv}) \tag{5.8}$$

On the basis of Berthelot's rule, Li and Neumann [7] introduced a modified combining rule:

$$\varepsilon_{ij} = \sqrt{\varepsilon_{ii} \varepsilon_{jj}} e^{-\beta(\varepsilon_{ii} - \varepsilon_{jj})^2} \tag{5.9}$$

where ε_{ij} is the energy parameter of the interaction between dissimilar particles, ε_{ii} and ε_{jj} are the energy parameters between similar particles, β is an empirical parameter, and the square of $\varepsilon_{ii} - \varepsilon_{jj}$ is used for taking into account the symmetry of the combining rule. Because the energy of adhesion (W) is proportional to the energy parameter, Eq. 5.9 can be written as follows:

$$W_{sl} = \sqrt{W_{ll}W_{ss}}e^{-\beta(W_{ll}-W_{ss})^2} \tag{5.10}$$

Thermodynamically expressed, the relation of the energy of adhesion per unit area of a solid–liquid pair is equal to the work required to separate the unit area of the solid–liquid interface:

$$W_{sl} = \gamma_{lv} + \gamma_{sv} - \gamma_{sl} \tag{5.11}$$

With the definitions $W_{ll} = 2\gamma_{lv}$ and $W_{ss} = 2\gamma_{sv}$ and a combination of Eqs. 5.10 and 5.11, the equation-of-state can be written as follows:

$$\gamma_{sl} = \gamma_{lv} + \gamma_{sv} - 2\sqrt{\gamma_{lv}\gamma_{sv}}e^{-\beta(\gamma_{lv}-\gamma_{sv})^2} \tag{5.12}$$

Combining Eq. 5.12 with Eq. 5.6 gives

$$\cos\theta = -1 + 2\sqrt{\frac{\gamma_{sv}}{\gamma_{lv}}}e^{-\beta(\gamma_{lv}-\gamma_{sv})^2} \tag{5.13}$$

The surface free energy (γ_{sv}) obtained from the model depends on the experimental constant β. The values γ_{sv} and β can be determined by a least-squares analysis technique from the set of γ_{lv} and θ data obtained with different liquids on the same solid. Spelt and Li [8] determined the value of β to be 0.0001247 (m^2/mJ)^2 from the data of measured contact angles with various liquids on three different surfaces. Recently, measurements with various liquids and polymer surfaces have yielded β values close to this value; this has resulted in a minor difference of approximately 0.1 mJ/m^2 when γ_{sv} is calculated [13]. The selection of liquids used by the "equation-of-state"—community has been unfortunately rather limited. They have used combination of liquids and solids that make the system essentially apolar. Thus, Lewis acid–base interactions are not revealed with this kind of experimental set-up. Even in the cases when polar substrates have been used together with polar liquids the combination has always been such that both solid and liquid have been monopolar in the same sense (i.e. either both are Lewis acid or Lewis base). This means that also in this case polar (actually Lewis acid–base) interactions are not revealed (see discussion below) and these liquids behave as apolar liquids on top of apolar solids. Therefore, the equation-of-state model seems to be very limited in its applicability. Finally, the model contains some elements that cannot be regarded as correct. It basically attempts to prove that all liquids having the same surface tension will have the same contact angle on an arbitrary solid. In other words, the interfacial tension between any two phases having the same surface tension will always be the same. Several experimental as well as theoretical investigations have been carried out to test the proposed equation-of-state model, but the model has failed to meet the tests [18, 19]. In fact the assumptions inherent in the model violate the Gibbs phase rule, thus making the model incompatible with thermodynamics. Attempts to justify the model by utilising the extended phase rule for capillary systems are not valid either. Laplace pressures start to influence the thermodynamics of wetting systems only when the radius of the drops starts to approach a fraction of micrometers. With such dimensions one can hardly talk about observable contact angle anymore.

5.2.2 Surface Free Energy Component Models

Fowkes [14, 15] adopted a different approach by assuming that the attractive forces between the surface layers and liquid phase across the interface were independent of each other, additive and contributed to the surface free energy. This approach was based on the earlier work of Girifalco and Good [5, 6]. Fowkes defined the dispersion force interaction between the solid and liquid phases and formulated their contribution to the surface free energy:

$$\gamma_{lv}(1 + \cos\theta) = 2(\gamma_s^d \gamma_{lv}^d)^{1/2} \quad (5.14)$$

This enabled the evaluation of the dispersion component (γ_d^s) of the surface free energy with liquids for which only dispersion forces operate like hydrocarbon liquids, but not the total γ_s.

Owens and Wendt [16] and Kaelble and Uy [17] extended Fowkes' equation to a geometric mean model that enabled the resolving of dispersion (γ_d^s) and polar (γ_p^s) components of the surface free energy:

$$\gamma_{lv}(1 + \cos\theta) = 2(\gamma_s^d \gamma_{lv}^d)^{1/2} + 2(\gamma_s^p \gamma_{lv}^p)^{1/2} \quad (5.15)$$

The value (γ_d^s) refers to dispersion (nonpolar) forces, and (γ_p^s) is assumed to refer to all polar (nondispersion) forces between the solid and liquid phases, including dipole–dipole, dipole-induced dipole, and hydrogen-bonding forces. The surface free energy is assumed to be a sum of (γ_d^s) and (γ_p^s) components, and so Eq. 5.15 provides a model for evaluating γ_s of solids with two liquids, one polar and one apolar, with known surface tension components (γ_d^{lv} and γ_p^{lv}). Because (γ_d^s) and (γ_p^s) are sensitive to surface chemistry, it was also proposed that the model could be used as a qualitative measure of the chemical composition of a surface. It should be noted that Eq. 5.15 also contains some questionable assumptions. It is not obvious that polar terms in Eq. 5.15 can be combined in Bertelot-type relations, which were originally developed for dispersion interactions between two different gas molecules. Another question is whether the solid surface tension is simply the sum of the interactive terms. One should be aware that van der Waals interactions among groups of molecules are not simply a sum of the individual pairwise interactions because they are affected by the presence of other interacting bodies in the vicinity [30]. The same assumption is also included in the following approach of Wu [20].

Wu [20] also assumed interfacial forces to be additive and proposed a harmonic mean model based on empirical observations:

$$\gamma_{lv}(1 + \cos\theta) = \frac{4\gamma_s^d \gamma_{lv}^d}{\gamma_s^d + \gamma_{lv}^d} + \frac{4\gamma_s^p \gamma_{lv}^p}{\gamma_s^p + \gamma_{lv}^p} \qquad (5.16)$$

Wu claimed that this model predicted the surface free energy components accurately in polar–polar systems [20].

Continuing the work of Fowkes [14, 15] and Owens and Wendt [16], van Oss et al. [18, 19] proposed the three-liquid acid–base model. It had been noticed that not all the interactions could be explained by simple division into dispersion and polar components, and the acid–base attraction between materials was expected to explain such cases [19]. In this model, γ_s is assumed to be the sum of a Lifshitz/van der Waals component γ^{LW} (corresponding to γ^d) and a Lewis acid–base component γ^{AB} (corresponding to γ^p):

$$\gamma_s = \gamma^{LW} + \gamma^{AB} \qquad (5.17)$$

The value γ^{AB} results from electron–acceptor (γ^+) and electron-donor (γ^-) Lewis acid–base interactions. The γ^{AB} term is expressed as a product of the electron-donor and electron-acceptor parameters,

$$\gamma^{AB} = 2\sqrt{\gamma^+ \gamma^-} \qquad (5.18)$$

Finally, the model was formulated by van Oss et al. [16]:

$$\gamma_{lv}(1 + \cos\theta) = 2\left[(\gamma_s^{LW}\gamma_{lv}^{LW})^{1/2} + (\gamma_s^+ \gamma_{lv}^-)^{1/2} + (\gamma_s^- \gamma_{lv}^+)^{1/2}\right] \qquad (5.19)$$

The value γ_s can be evaluated by Eq. 5.19 being solved simultaneously for three liquids (one apolar and two polar) used in the measurement of the contact angle with known surface tension components. A material is considered bipolar if both its γ^+ and γ^- components are greater than 0 ($\gamma^{AB} \neq 0$). A monopolar material has either an acidic or base nature; that is, either $\gamma^+ = 0$ and $\gamma^- > 0$ or $\gamma^- = 0$ and $\gamma^+ > 0$. An apolar material is neither an acid ($\gamma^+ = 0$) nor a base ($\gamma^- = 0$). It should be noted that there are many polar solids and liquids that are strong electron donors but have little or no electron acceptor capacity [31]. The opposite situation is also possible, but is rather rare [32]. Thus, these compounds have a strong Lewis acid or base character, but in the absence of a surface tension parameter of the opposite sign these do not contribute to the energy of adhesion. However, when these monopolar compounds are contacted with bipolar liquids, for example water, they can strongly interact with the opposing polar parameter (γ^+ or γ^-) in the bipolar substance. There are, in principle, seven possible classes of binary systems regarding the interactions discussed above. These are schematically shown in Fig. 5.3.

In this figure, LW interactions are shown by solid arrows and AB interactions by interrupted lines. Horizontal arrows are indicative of cohesive and vertical

5.2 Surface Energy and the Wetting of a Solid Polymer Surface

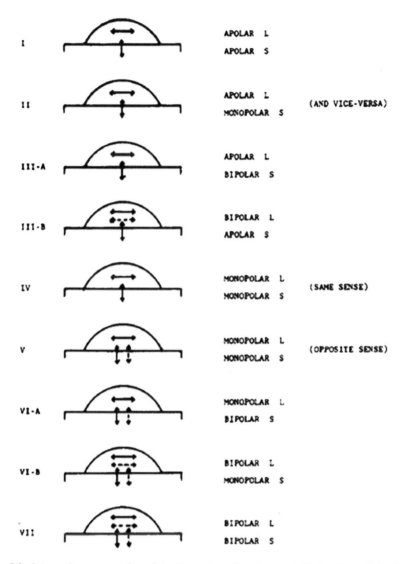

Fig. 5.3 Schematic representation of the interactions that play a role in the shape of the liquid droplets on solid surfaces. LW interactions are shown by arrows connected by solid lines and the AB interactions by interrupted arrows [18]

arrows of adhesive interactions. Cohesive interactions have not been marked in solid as they typically do not appreciably contribute to the shape of the liquid drop. On the other hand, cohesion in the liquid forming the drop generally plays an important role in determining the final shape of the liquid drop. This is owing to the competition between the adhesive forces (towards solid) that tend to flatten out the drop and the cohesive forces in the liquid that try to revert it to a spherical shape.

The surface free energy of a given solid is by no means a simple matter. It cannot be assumed to be a material constant because it is crucially dependent on the history of the sample, the surface reconstruction, the cleavage plane, and so forth. Therefore, it is our opinion that the surface free energy values of solids should be taken as indicative and not as explicit properties and only to give some indication about the property of the particular solid surface in question with a known history. The techniques discussed above will provide experimental means to evaluate and not to determine surface energies of solids. However, evaluation of surface energies of treated surfaces is a valuable tool to follow up the changes in the state of the surface and to understand the complicated phenomenon of adhesion at least to some extent. In Chap. 6 these methods have been utilised and several examples are provided for polymer cases.

5.3 Kinetics of Wetting

So far, wetting has been considered from the viewpoint of thermodynamics. Although the thermodynamics may indicate possible formation of an intimate contact between two phases, the kinetics will more likely be the determining factor for the occurrence of the wetting. We now explore the kinetics of wetting.

5.3.1 Role of Surface Roughness and Capillary Action

Roughness will affect the contact angle (θ) of a given liquid on a solid. If the equilibrium contact angle on a smooth surface is less than 90°, roughening the surface will make θ even smaller. This will increase the apparent surface free energy of the solid and thus also the extent of wetting. However, if for a smooth surface θ is greater than 90°, roughening the surface will increase θ and retard wetting. For spontaneous wetting to take place, the equilibrium contact angle of a liquid on a given smooth surface has to approach zero. Spreading of a liquid on a solid surface may be accelerated by capillary action. Capillary forces effectively utilise fine pores, scratches and other topographical inhomogeneities when a liquid drop forms a contact angle less than 90° [33–35].

Wenzel [36] formulated the relationship between the roughness, thermodynamic contact angle θ and apparent contact angle θ_{app} in Eq. 5.20 by employing a roughness factor, r:

$$A = rA_{smooth}; \quad r = A/A_{smooth} > 1 \qquad (5.20)$$

5.3 Kinetics of Wetting

This was further expressed in the modified form of Young's equation which is known as Wenzel's equation.

$$\cos \theta_{app} = r(\gamma_S - \gamma_{SL})/\gamma_L \qquad (5.21)$$

This equation can be used to assess the effect of roughening a given surface on the apparent (thermodynamically preferred) contact angle. We see from the above given results that the wettability of a surface that already wet ($\theta < 90°$) is increased by roughening and the wettability of a surface that is not wet ($\theta > 90°$) is decreased. Since r > 1 for a roughened surface, if cos θ is positive ($\theta < 90°$), the roughened surface will have a larger cos θ_{app} (Eq. 5.22) and vice versa in the case of nonwetting for which cos θ_{app} will appear more negative when cos θ is negative ($\theta > 90°$).

$$\cos \theta_{app} = r \cos \theta \qquad (5.22)$$

The boosting effect on wettability by roughening is important in many cases. In particular, if θ_{app} can be lowered to zero degree, spreading will occur and effective utilisation of the cavities of porous surfaces will be achieved which in turn is important (at least) in cases where the adhesion mechanism is based on mechanical interlocking between a porous substrate and a coating applied in a liquid state like many lithographic polymers, adhesives, etc.

5.3.2 Role of Chemical Surface Homogeneity

A chemically heterogeneous surface contains domains of different surface free energy and obviously the intrinsic contact angles for these domains are different θ_1 and θ_2, respectively. The cosine of apparent contact angle θ_{app} on a heterogeneous surface can be expressed by the following Cassie-Baxter equation after the minimisation of the system free energy [37]

$$\cos \theta_{app} = f_1 \cos \theta_1 + f_2 \cos \theta_2 \qquad (5.23)$$

where f_1 and f_2 are area fractions for the two domains of the area. The equation can be extended to one very important case in which pores in the surface lead to vapour gaps across which the liquid does not contact the solid. The θ_2 over such gaps is 180°, and the Cassie-Baxter equation becomes:

$$\cos \theta_{app} = f_1 \cos \theta_1 + f_2 \qquad (5.24)$$

If the heterogeneous domains are very small relative to the size of liquid drop used in the measurement of the contact angle, which is common, it will be difficult to differentiate the influence of the different domains on the contact angle [38]. Even microscale surface heterogeneity has been observed, however, through the use of a special micro droplet condensation technique [39]. In practice, the chemical heterogeneity may result in compromised wetting, varying interfacial adhesion, difficulty in measuring contact angles reproducibly and, in particular, severe contact angle hysteresis (i.e. the difference between an advancing and a receding contact angle). It has been proposed that the contact angle hysteresis originates from chemical and topographical surface heterogeneity. Many processes taking place on the surface also have an effect on the contact angle, such as adsorption and desorption of the probing liquid, reorientation of functional groups, surface deformation and surface contamination [40–43].

Although homogeneity of the surface has so far been the desired state, controlled surface heterogeneity may also be of use. Spontaneous orientation and redirection of functional groups can be achieved by applying an external control [44–46]. Orientated and selectively modified surfaces have been utilised to control both the thermodynamics and kinetics of wetting, for example, to enable the build up of nanoscale optical wave guides and fluidic microsystems [47–52].

5.3.3 Influence of Interphasial Interactions

Despite the apparent simplicity, the spreading of liquid on solid surfaces may involve many complex physical processes including swelling, dissolution, diffusion and reactions. For a solid polymer surface, it is fairly difficult to study processes like reactive wetting. In metallurgy, in contrast, a great deal has been published on the physical processes that affect the kinetics of wetting [53–55]. An example relevant to electronics is soldering, where the oxidised copper substrate is first reduced by fluxes and then the exposed metal is reactively wetted by a liquid solder that simultaneously engages in mutual dissolution and interdiffusion with the surface [56].

5.3.4 Influence of liquid viscosity

Viscosity describes the flow behaviour of a material. It originates from intermolecular friction, and so hinders the spontaneous spreading of liquid over solid surfaces. For adequate wetting, liquids should possess relatively low viscosity at some stage of the processing. Often the intrinsic viscosity is too high and macromolecules are diluted in solvents for viscosity reduction before their use in coating. The use of elevated process temperature is another effective way to enhance wettability. Increased temperature enhances molecular mobility, separating molecules apart from one another, which in turn decreases both the viscosity and surface tension of the liquid. Gastellan and Defay et al. [57, 58] Nevertheless, thermosetting polymers such as those used as encapsulation and underfill materials start to crosslink during thermal processing, which rapidly increases their viscosity and complicates spreadability. Wang [59] Industrial adhesives are often so heavily

5.3 Kinetics of Wetting   115

filled with inorganic additives that the viscosity (kinetics) and even wettability (thermodynamics) is severely compromised. This is adversely seen in the air bubble formation at the interfaces of polymer/silicone adhesive and chip/silicone adhesive. Needless to say, this ultimately compromises the integrity of the adhesion and thus the reliability of electronic components.

The spreading kinetics of liquids on solid substrates has been studied simply by monitoring the areal coverage as a function of time [60] however, theoretical models have been proposed by de Gennes [53], Tanner [61], and Seaver and Berg [62]. The models assume that the most significant forces affecting spreading are the liquid's surface tension, which is the driving force, and viscosity, which is the counterforce. The models have been tested [60] and they hold for some cases but not all. In spite of the practical importance of understanding surface phenomena and the theoretical foundation laid by Young and Laplace almost 200 years ago [58], the surface processes are still poorly understood [63].

5.4 Surface Modification

This section deals mostly with the surface modification of polymers. Metal and ceramic modifications are only briefly reviewed and only in cases where they are to be coated with a polymer. Metal–metal, metal-ceramic and ceramic–ceramic systems are excluded.

Most polymers have low surface free energy and are not amenable to interfacial adhesion without a change in their inherent surface chemistry. Modification of polymer surfaces can be carried out without altering the bulk properties and materials mechanical integrity. However, the affected layer, i.e. the interphase will then differ significantly, in both physical and chemical properties, from the bulk material. Often the roughness of the surface is increased as well. The thickness of the affected layer varies with the material, but usually the chemical modification reaches the first few nanometres and if significant roughness is produced it may reach even a few micrometers. Although many techniques are available for polymer surface modification, they can effectively be divided into the two major categories of wet and dry processing. In addition, there are various chemical tricks to link coatings onto a substrate or fillers to a polymer matrix. Adhesive promoters are applied either on the substrate or in the bulk material of the coating. The promoter molecule may act as a compatibiliser that increases wettability (non reactive interactions) or it may perform as a coupling agent that reacts with both the substrate's and coating's functional groups. The use of compatibiliser allows adhesion between two materials that inherently repel others.

In wet processes, polymers are immersed in chemical solutions, where as in dry processes they are subjected to vapour-phase species. Dry processing basically involves a bombardment from above the sample, during which the reactive species travel along the direction of an electric field established between the electrodes above and beneath the sample. Unlike the wet processing, this results in a

Table 5.1 Effects of the surface treatments on the root mean square (rms) roughness of unfilled and filled polymers [64–68]

Type of epoxy	Untreated (nm)	Wet-chemically treated (nm)	RIE treated (nm)
Unfilled epoxy	0.3 (smooth)	2.0 (smooth)	12.7 (spiky)
Particle filled epoxy	4.0 (smooth)	300–1200 (porous)	10.0 (wavy, crater)

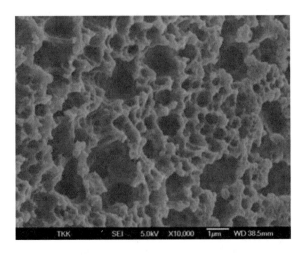

Fig. 5.4 SEM micrograph of a wet-chemically treated photodefinable epoxy. Specimen tilted 45° to obtain best three dimensional impression of the porous surface [69]

mountain-like topography without significant creation of cavities. Some micrographs of the treated surfaces are shown in Chap. 6 together with the polymer cases. Table 5.1 shows schematically the different outcomes with wet-chemical and RIE treatments of unfilled and filled polymers.

Wet processing is composed of three stages: swelling, etching and neutralising. The purpose of the swelling is to make the topmost surface layer susceptible to the etching solution. The time used for swelling determines the depth of the resulting modification. The swelling solution diffuses into the surface layer and softens it by plasticisation. The $KMnO_4$ component of the etching solution will then readily diffuse into the swelled layer and attack the material through oxidation of the polymer molecules. Oxidation breaks the polymer molecules into small-molecular weight fragments, which dissolve in the solution. Oxygen is often also incorporated into the remaining chains through the creation of new functional groups like hydroxyl and carboxyl. If, and only if, there is more than one phase in the polymer material, a microporous surface can be achieved. See Fig. 5.4 for a wet-chemically treated epoxy.

The second phase in a polymer material may be another polymer that is more readily etched than the primary matrix polymer. This sacrificial phase may form a

5.4 Surface Modification

continuous or separate phase within the primary matrix but it may not be fully soluble if a microporous surface layer is aimed for from the wet processing. Often the second phase is an inorganic filler like calcium carbonate, glass fibres, talc, or a flame retardants such as ATH (Aluminumtrihydrate) or $Mg(OH)_2$ (Magnesiumhydroxide). There are many reasons for adding fillers including lowering the cost of the polymer ($CaCO_3$, Talc), making it mechanically tougher or stiffer (glass fibers, $CaCO_3$) or making it flame retardant to prevent burning when ignited (ATH, $Mg(OH)_2$). The fillers may be present up to very high levels (80%). To produce a microporous surface, only a small amounts of filler material is needed and the roughening mechanism is based the dropping of these filler particles from the surface [64–68].

The oxidative treatment can also be applied without prior swelling to obtain a plain chemical modification of the surface with minimized roughening effect. Chemically homogeneous polymers like Epon SU-8 do not show any roughening effect if etched with wet-chemical of dry processes. Only chemical oxidation of the surface can be detected. This provides a poor means basis for adhesion of copper or most other metallisation.

The use of swelling solution accelerates the etching rate even with homogeneous polymers, but with heterogeneous polymers an appropriate swelling treatment ensures markedly different etching rates in the chemically different phases providing that there is a difference in their chemical stability and the solubility of the sweller in the less stabile phase. It needs to be mentioned that prolonging the etching treatment endlessly does not have an effect on the modification depth or the degree of oxidation. This is because the redox reaction that takes place during etching yields MnO_2 deposits on the polymer, which effectively inhibit diffusion of the reactive $KMnO_4$ into the polymer and thus prevent the oxidative etch-reaction from continuing. This MnO_2 layer is finally removed in the neutralising solution [70, 71].

Different polymer chemistries require different etching solutions. Many etching solution recipes are presented in literature. On a commercial polyimide (PI), chemical modification of the surface together with the roughening may have an enhancing effect on the adhesion. According to the literature, the imide ring of a polyimide substrate is opened in an alkaline bath [72]. The polyimide surface reacts in a water-based NaOH solution to produce a sodium salt of polyamic acid, i.e. polyamate. This polyamate in turn, can be protonated with an acid to give a surface rich in polyamic acid [72, 73]. The polyamic acid surface has been reported to provide good adhesion between successive polyimide layers [72], and the incorporation of hydrophilic groups on the polymer surface has been reported to yield good adhesion with sputter-deposited copper [72–74]. In the case of a polyimide substrate and electrolessly deposited copper, microroughening of the surface is also required [75].

Dry processing etches surfaces through the use of chemically reactive gases or through a physical bombardment of gaseous atoms that resembles sandblasting. Reactive ion etching (RIE) involves both chemically reactive species and physical bombardment to provide anisotropic etching with a fast rate, whereas plasma etching is based on the use of chemically reactive species alone [76, 77]. Reactive species are absorbed onto the surface where they form volatile etch products that are eventually desorbed from the surface and removed by a vacuum pump. Since the energies of ions are relatively low, etch rates in the vertical and lateral directions are often similar, resulting in an isotropic etching unless there are several phases in the treated material. Homogeneous polymers remain reasonably smooth regardless of the method used in their modification. Polymer blends in which reaction induced phase separation occurs during a crosslinking bake are particularly prone to obtain a microporous surface during treatment if one of the separated phases is chemically much weaker towards the etching medium than the other(s). The main consequences of plasma exposure are the removal of organic material and the oxidation of the remaining polymer surface. The same takes place in RIE, but there ion bombardment is usually dominant over the chemical etching and the resulting surface is topographically changed rather than chemically altered.

5.5 A Brief Look at the Adhesion Mechanisms

Typically a number of different mechanisms for adhesion are listed and the list includes: (i) the adsorption mechanism, (ii) the diffusion mechanism, (iii) the mechanical interlocking mechanism and (iv) the electrostatic mechanism. As hardly any of the mechanisms listed above work alone in real adhesion cases, practical adhesion becomes rather a combination of contributions form different mechanisms.

5.5.1 Adsorption Mechanism (Contact Adhesion)

In the adsorption mechanism the adhesion is taken to occur across a well defined interface by molecular interactions across that interface. Owing to the relatively short range nature of molecular interactions the adsorption mechanism is restricted to the vicinity of surfaces of the adhesive and the adherent. The forces interacting across the surface may be secondary (i.e. strength below 100 kJ/mol) or primary (strength above 100 kJ/mol). Secondary forces involve mainly Lifshitz-van der Waals interactions, which include the dispersion forces (London force), dipole–dipole forces (Keesom) and dipole-induced dipole forces (Debye). Typically, only dispersion forces are of importance (magnitude is about 0.1–10 kJ/mol). In addition to the universally present LW interactions, there may be an acid–base, i.e. donor–acceptor interactions. These occur when an electron donor (Lewis base) shares some of its electron density with an electron acceptor (Lewis acid) to form a acid–base complex. These interactions include also hydrogen bonding (magnitude is about 10–25 kJ/mol). There is also a possibility to form primary chemical bonds, i.e. covalent, ionic or even metallic bonds, during adhesion event. The two

5.5 A Brief Look at the Adhesion Mechanisms

most important factors determining the interfacial strength in contact adhesion are (i) the completeness and intimacy of the contact between the materials at the interface and (ii) the strength of the intermolecular interactions across the interface. These are also the factors that one can attempt to modify as shown in Chap. 6.

5.5.2 Diffusion Mechanism

In many cases when polymers are bonded together (and especially in the case of bonding identical polymers together) interdiffusion across the contact interface is observed. This takes place especially if the solubility between the polymers is adequate, the polymer molecules are mobile enough and there is sufficient contact time. As a result of this interdiffusion, formation of an interphase takes place. The thickness of this interphase may be of several micrometers but typically smaller values are encountered [77]. Initial intimacy of the contact between the materials in diffusion mechanism is of course required, but in contrast to the adsorption mechanism the parameter to optimise is not the interfacial interactions but the phase compatibility in the form of mutual solubility.

5.5.3 Mechanical Interlocking Mechanism

The higher interfacial strengths often achieved by roughening the adherent surface are often explained with the help of increased mechanical retention. It is thought that especially the liquid adhesive can penetrate the cavities in a rough adherent surface resulting in mechanically effective hooks that hold the materials together. Aside of any hooking effects, the roughening increases the surface area across which the interfacial forces act and may induce microstructural changes to the underlying material. Perhaps most importantly the roughness seems to increase the energy needed to fracture the contact interface by diverting the fracture forces away from the interface into the bulk of the materials.

5.5.4 Electrostatic Mechanism

Almost always when interfaces are formed there is a electrical double layer present. The double layers are always charged (i.e. there is potential difference across the interface) and it has been suggested that electrostatic interaction would also contribute to adhesion. However, this contribution is generally considered to be small [78, 79].

5.5.5 Adhesion Mechanisms in Practice

When two metals or ceramics are joined together in a high temperature process like soldering/sintering they often react to form an alloy or diffuse into each other's forming solutions at the interface, or, more precisely, at the interphase that adheres those inorganics well in a chemical sense. However, when polymers that may appear chemically quite inert, are joined at considerably lower temperatures than inorganic materials, both diffusion and reactivity are limited. The lack of diffusion and reactivity are particular problems in cases that a polymer substrate is coated with another polymer layer or with inorganic materials. Obviously in cases that inorganic substrates are coated with reactive thermosetting polymers some covalent or ionic bonding is probable as such (and particularly when coupling agents or adhesion promoter layers are employed). Important thing here to realise is the role of the sequence in which different materials are brought into contact on the practical adhesion strength and mechanisms.

Table 5.2 presents the rule of thumb to the mechanisms that can be expected at least to take place in model interfaces of similar and/or dissimilar materials. The information in the table may be used to deselect material combinations that will most obviously result in adhesion problems. The information in the table also provides means to select those adhesion mechanisms with which one should work when choosing a surface treatment method in order to influence and improve the adhesion.

Discussion on the key adhesion mechanisms in polymer substrate—metallization—systems

Mechanical Interlocking provides improved interfacial adhesion by two mechanisms. First, electroless metal deposition may utilise the fine pores prepared on the surface of a polymer and thus anchor into them. Second, if anchoring is not possible the rough surface will still dissipate the force applied to the interface and thus more force is actually needed to break the interface resulting in higher apparent adhesion strength. Diffusion of high molecular-weight polymers into metals is not reported. Metals do not spontaneously diffuse into polymers in normal application conditions of electronics. But, during the deposition process some metals may diffuse into the polymer substrate sufficiently to provide improved adhesion, if metal atoms are brought onto the polymer substrate with sufficient kinetic energy (e.g. sputter depostition). Primary Bonding or chemisorptions between polymer and metal has been reported in some cases in the form of metal complex formation (for further discussion see polymer cases in Chap. 6). The study of these complexes is extremely difficult in the interface. They could, however, partly explain high adhesion strengths and the shift from adhesion fractures to cohesive ones in polymer-metal systems. Complex formation may be enhanced by surface modifications of the low surface energy polymers by forming polar functional groups on the surface [80–84]. Secondary Bonding or physisorption between polymer and metal can result in good adhesion performance if good contact between polymer and metallisation is achieved. But, this may

5.5 A Brief Look at the Adhesion Mechanisms

Table 5.2 Adhesion mechanisms at different interfaces

Material type deposited/coated on another material type	Adsorption	Diffusion	Mechanical	Electrostatic
Thermoplastic on thermoplastic	Secondary forces dominate	Highly possible when solubility is high	Possible	Possible, if poor solubility prevents diffusion
Thermoplastic on thermoset	Secondary forces	Unlikely	Possible	Highly likely
Thermoset on thermoplastic	Secondary and primary bonds possible	Possible when high process temp. and solubility	Possible	Highly likely
Thermoplastic on ceramic	Possible	Not possible	Possible	Likely if polymer wets the substrate
Thermoset on ceramic	Possible	Not possible	Possible, porous ceramics	Possible
Thermoplastic on metal	Possible	Not possible	Possible	Possible
Thermoset on metal	Highly possible	Not possible	Possible	Highly likely
Metal on thermoplastic	Possible	Possible but highly depended on the metallization method	Possible, depends on deposition method	Possible, depends on deposition method
Metal on thermoset	Possible, not likely	Possible but highly depended on the metallization method	Possible, depends on deposition method	Possible, depends on deposition method
Ceramic on thermoplastic	Unlikely	Possible, depends on deposition method	Unlikely	Possible
Ceramic on thermoset	Unlikely	Unlikely	Unlikely	Unlikely
Metal on ceramic	Possible	Possible in high process temps.	Possible, depends on deposition method	Possible, depends on deposition method
Ceramic on metal	Possible	Possible in high process temps.	Unlikely	Possible
Metal on metal	Highly likely	Highly likely if solubility is good	Possible, depends on deposition method	Likely
Ceramic on ceramic	Possible, depends on deposition method	Possible, depends on deposition method	Unlikely	Possible

demand the application of external force especially when metal films are laminated.

Discussion on the key adhesion mechanisms in metal substrate—polymer coating/adhesive—systems

Mechanical Interlocking provides the same benefits that were discussed above. Diffusion of polymer coating into the metal substrate is not possible. Primary Bonding of polymer coating to the metal is possible. Still, often the use of coupling agents or adhesion promoters is needed to form a stable chemical link between a metal and a polymer. These surface active low-molecular-weight additives are applied on the metal substrate as very thin monomolecular layers from dilute solutions. The number of links is high per area and in some cases these linking molecules may further react to form a crosslinked coating on the metal and thus stabilise the system before coupling to the polymer coating. By pointing outwards from the metal surface some polar or other reactive groups like carbon–carbon double bonds the applied polymer coatings, providing that they are reactive towards the functional groups, may link to the coupling molecules. Secondary Bonding is very likely to appear when polymers are applied on metal surfaces. Metals, particularly reduced surfaces, poses high surface energy and thus provide adequate means for good wetting. Nevertheless, oxidised metal surfaces may provide better adhesion than reduced metals surfaces through chemical bonding, see Chap. 6.

Discussion on the key adhesion mechanisms in polymer—polymer—systems

Mechanical Interlocking contributes to improved adhesion if the surface free energy of a polymer substrate is not too low. Of equal importance is to bear in mind the old fact that polar surfaces are more easily wet by polar polymers, hydrophobic surfaces like hydrophobic polymer coating materials, and thus polar prepolymers are repelled by hydrophobic polymers surfaces. Diffusion of thermoplastic polymer molecules across thermoplastic polymer–polymer interface is possible in certain cases providing that the polymers are mutually soluble—rarely in the case of thermosets. The process temperature has to exceed the glass transition temperature of both substrate material and coating material for sufficient time. Also in those cases where coating is applied from such solutions that dissolve the substrate material diffusion is possible. Often, however, the polymers used in electronics are either crosslinked thermosets or chemically inert thermoplastics like liquid crystal polymers (LCP) that do not inherently exhibit mutual solubility or there are no kinetic prerequisites for diffusion to take place. Primary Bonding is often the most feasible mean to ensure good interfacial adhesion and reliability of the interface when polymer–polymer systems are exposed to environmental stresses. Thermosetting polymers may be intentionally left partly uncrosslinked (prepregs) and to some extent reactive functional groups may be produced on the thermoplastic surfaces by proper treatments or they may contain such groups inherently. The covalent bonds across a polymer–polymer interface are typically carbon–carbon or silicon-based and thus water repellant providing reliable performance in various environments. Secondary Bonding contributes greatly to the interfacial adhesion at the polymer–polymer interface especially if the substrate is

5.5 A Brief Look at the Adhesion Mechanisms   123

adequately pretreated. Many thermosetting prepolymers are polar in nature and thus oxidative plasma or wet-chemical treatments are often utilised for the substrate to which the prepolymer is to be applied. If the interfacial adhesion is mainly a result of secondary bonding the polar interfaces are susceptible to interfacial diffusion of water in moist environments. This diffused water will act as a plasticiser or lubricant in the interface and thus effectively reducing the adhesion strength. In some cases hydrophobic coatings like polyethyle (PE) are used and obviously the substrates to be coated should appear hydrophobic also. These interfaces are less prone to the adverse effects of water if exposed to moist environments.

5.6 Durability of Interfacial Adhesion

So far, the discussion has focused on the essential role of interfacial adhesion in the manufacture of reliable electronic devices. Of equally great concern is the durability of the adhesion in the changing and often harsh environments to which consumer products are exposed. Destructive processes like corrosion, deformation and fatigue are accelerated by environmental factors, and eventually failures in one form or another will occur. Materials with physically and chemically different properties are frequently combined in electronics, and the interfacial adhesions are thus particularly susceptible to changes in temperature, moisture ingress and mechanical shocks. A fundamental understanding of the interfacial compatibility of dissimilar materials is even more important for the miniature structures of advanced electronics where relatively short times and low stress levels are sufficient to cause problems.

The loss of bulk properties of polymers is rarely a reason for the failure of joints attacked by environmental factors. Although the locus of failure of well-prepared joints in adhesion tests is invariably a cohesive fracture in the polymer layer, the exposure of specimen to environmental attacks frequently results in an "apparent" interfacial failure with lowered adhesion strength [85]. This finding underlines the importance of the interface when considering environmental failure mechanisms.

Since polymers have a significantly higher portion of free volume in their structure than metals or ceramics, it is reasonable to expect that when the temperature of an environment in which multimaterial devices are used is changed, the polymer expands significantly more than other materials. If the temperature does cycle during the use, like it does in many everyday electronic devices either inherently (local heating of the device in use) or because of environmental factors, there will appear interfacial stress cycling between polymer (adhesives, coatings, sealing materials, etc.) and other materials. The bulk materials may relax during the cycling but the chemical bonds across the interface must stretch and if they cannot do that they must partially break causing gradual weakening of the adhesion and eventual delamination. Printed wiring boards with their high load of class fiber filling may perform in away somewhat similar to inorganic materials rather than to polymers containing less inorganic fillers.

With their high free volume, polymers are more prone to the diffusion of other materials and liquids than metals in normal operational temperatures. In high temperature and high moisture environments this ability of polymers to absorb water, for example, increases and may cause many degradation processes. First of all, water tends to swell polymers when it penetrates into the bulk material. This results in stress at the interface although it acts as a flexibiliser in the bulk as it reduces the intermolecular forces between polymer chains. This in turn reduces the mechanical properties of the bulk polymer material and makes it more vulnerable to other stresses. Water may eventually reach the interface and there it may attack on the covalent bonds and hydrolyse them if that is thermodynamically possible or the water may cause corrosion of metals at the interface and cause delamination as the volume of corrosion products increase. It is to be said that diffusion of water to the interface is more probable through interfacial diffusion than through bulk diffusion when proper polymer materials are used. But if the interfacial adhesion is inherently poor and if there are thermodynamic prerequisites for the penetration of water to the interface, it is likely that in moist environments delamination occurs seemingly fast. Degradation of the epoxy coatings in humid environments has been studied thoroughly by many authors [86–88] since both temperature and moisture changes alone and particularly simultaneously exposure of polymer structures to elevated temperatures and moist environments probably result in most problems rather than other stresses like mechanical shocks.

5.7 Measurement of Interfacial Adhesion

Before starting an adhesion test, it is imperative to gather as much information about the materials as possible. False interpretation is likely if careful consideration is not paid to factors like test geometry and the measurement technique and their suitability to the particular material.

The most common methods for measurement of interfacial adhesion can be divided into the two major categories of tensile and shear tests. However, for testing the adhesion of coatings to substrates, there are a vast number of variations of these basic tests including pull-off, lap joint and peel tests [89–97]. In addition, qualitative methods such as the tape test can be employed for quick evaluation of interfacial adhesion on a fail/pass criterion [95]. Indirect methods such as the cavitation test, have been employed to estimate the adhesion of TiN and similar hard films to substrates [93, 96]. It is recognised that the different methods do not provide comparable values of adhesion strength and sometimes they may even lead to contradictory results. The adhesion tests have been critically discussed in the literature [25, 85–90, 97].

The main reason for the selection of the pull-off or shear type of tests lays in the fact that the stress distribution within the specimens is well controlled. The pull-off test is a modification of the tensile test. However, every effort has to be made to ensure uniform stress distribution within the pull-off test specimen if the test is to be reliably applied. Shear-type tests, it may be argued, represent best the failure mechanisms found in field applications, but because of the more complex stress distribution within the specimen, the design of the test configuration has a pronounced effect on the adhesion strength value obtained [85]. Also the peel test is often used, but it is affected by several experimental parameters (e.g. angle, rate and width of the peel) as well as the bending of flexible foil and friction, all of which dissipate the energy intended to separate the two bodies of interest [89]. The lap shear and peel methods are usually used as comparative tests therefore whereas the pull-off method is used to obtain more comprehensive understanding of the interfacial adhesion.

There are basically only two adhesion failure modes (namely delamination at the interface and bulk fracture) in either of the materials or within their reaction product layer, i.e. interphase. If the adhesion is poor, the limiting factor or geometrical reasons concentrate effectively stresses on the interface, then delamination takes place. If the adhesion is good, the bulks' mechanical properties or geometrical reasons distribute the stresses away from the interface, the fracture is likely to take place some other place than at the interface. It will be discussed in the forthcoming text that adhesion strength may be calculated in the ideal bulk fracture cases where the thermodynamic work of adhesion and mechanical properties of the fracturing bulk material determine the fracture strength. However, in all the other cases where the failure of the interface is purely interfacial delamination or mixed mode of the series of break downs of different adhesion mechanisms, one must, first, gather as much interfacial data of the joint prior to joining and, second, collect fracture surface information after the destructive adhesion test using various surface analysis methods.

The topography, chemistry and surface free energy of the given surface should be characterised, first, prior to surface treatments, second, after the treatments before application of the coating and, finally, after the adhesion test (fracture surface where ever it then lays). The methods used in these studies are examined in more detail in Chap. 6.

In addition to destructive tests there are some means to follow the evolution of delamination, providing that the adhesion lose mechanism is the delamination, by in direct methods. Water penetration to the surface can be detected by capacitor structures [98]. Corrosion processes at the interface have been followed by, for instance, electrochemical impedance spectroscopy (EIS) [99–102]. EIS is also used in the quantification of interlayer adhesion in automotive coating systems exposed to different humidity conditions. Destructive methods are inconvenient to quantify time-dependent adhesion changes but EIS may provide means to predict the reliable lifetimes of coating systems. Also scanning acoustic microscopy (SAM) can be used but it demands immersion of the sample in the water and it may cause side effects like sudden delamination in very poor adhesion cases.

5.8 Work of Adhesion

In the absence of chemisorption and mechanical retention including diffusion, the thermodynamic work of adhesion (W) can be used to predict the durability of interfacial adhesion. Although the total absence of these forces at the interface of metal oxide and a thermosetting polymer is unlikely, the use of W may still help explain the weakening of interfacial adhesion during the environmental attack of liquids. The work of adhesion in air (W_a) and in the presence of liquid (W_{al}) is given in Eqs. 5.25 and 5.26, respectively.

$$W_a = \gamma_a + \gamma_b - \gamma_{ab} \tag{5.25}$$

$$W_{al} = \gamma_{al} + \gamma_{bl} - \gamma_{ab} \tag{5.26}$$

In Eqs. 5.25 and 5.26, the γ_a and γ_b are the surface free energies of the two phases in air (subscript l denotes the properties in liquid) and γ_{ab} is the interfacial free energy. W_a usually has a positive value indicating thermodynamic stability of the interface. In the presence of liquid, however, the thermodynamic work of adhesion, W_{al}, may well have a negative value indicating that the interface is unstable and will dissociate. Indeed, the thermodynamic work of adhesion at metal oxide/epoxy interfaces changes from positive to negative in the presence of water [85, 103]. Evidently a significant amount of primary bonds thus must have existed at the copper/epoxy interface studied in [98] because the epoxy did not delaminate during exposure to the atmosphere at elevated humidity and temperature. These epoxy coatings were prepared on a relatively smooth CuO_2 substrate and only minor statistically insignificant weakening was observed. In the case of a liquid crystal polymer (LCP) and silicone interface, we observed total loss of adhesion when exposed to a high humidity environment [104]. Moreover, the adhesion was regained after dehydration. In a previous study the water penetration and removal from the interface was monitored using interdigital capacitors [98]. For these interfaces that bear only secondary interactions, a faster method to introduce water to the interface may be utilised and it may take the form of pressure cooking [104]. In interfaces bearing hydrolysable primary bonds one should not over accelerate the environmental testing but adhere to the use of standardised test protocols like 85°C/85RH% for a given time.

5.8.1 Testing the Adhesion Strength-Quantitative versus Qualitative

Previously some basics of interfacial adhesion were discussed. If the principles of the adhesive interactions are well known, could they be modelled using modern simulation tools? It has been shown [105] that to detach two ideally connected bodies, there is a relationship for the work of adhesion between the two phases

5.8 Work of Adhesion

given in Eq. 5.20 but in practice [106] the adhesion forces are significantly less than they are theoretically expected due to imperfect bonding, etc. The influence of mechanical properties and geometrical effects on the practical adhesion has been demonstrated as early as 1920 by Griffith [107]. This can be good or bad for the adhesion. If we intentionally roughen the surface prior to bonding we may distribute the forces affecting the interface and reach better adhesion results that would be achieved with smooth surfaces. On the other hand, if we design the interface geometry of two materials in an improper way and end up concentrating forces at the interface or at some weak spot of the device, the resulting adhesion performance may be unexpectedly poor.

Now, having said that there is no direct relationship between the work of adhesion (W) and fracture energy (G) needed to separate two materials, there must, however, exist some parameters connecting the W and G. Andrews and Kinloch [108] showed that there exists a trend according to which the fracture energy G correlates with the work of adhesion W by

$$G = W \times \phi \tag{5.27}$$

where ϕ is a temperature and rate dependent viscoelastic term. This is based on fracture mechanics and may describe generally the action of breaking an existing joint. What it does not take into account is the circumstances at the moment of creation of the interface (viscosity, temperature, pressure, reactivity, inhomogenities, etc.) and hence increases little if at all the understanding of the adhesion mechanisms behind the simple fracture adhesion value. Actually, this relationship holds only for very simple systems [109–112].

After discussion all the above factors that are very difficult to mathematise ideally, the modelling approach may not be a feasible method today for the interpretation of adhesion strengths. It is, nevertheless, an effective tool to identify stress concentrations in the devices and it may be used to evaluate adhesion test geometries. In the first case obviously aiming at improving the reliability of the device and in the second case understanding better the results gained from the modified adhesion tests. In other words, real interfaces are unideal and there are too many unknown variables that may be too difficult to incorporate in a simple finite element modelling simulation. Therefore, we still have to carry out experimental tests to understand and develop the adhesion performance of components manufactured from dissimilar materials.

A good understanding of the interfacial adhesion forces can be obtained from a tensile type of test. The extreme tensile test would appear as the force measurement of the interactions between an atomic force microscope (AFM) tip and the studied surface. However, in nanoscale measurements we lack the engineering type of information gained from macroscale samples with geometrical, material property and other factors. Thus, the ideal way of testing adhesion strength between two engineering materials would be a tensile test that may take the form of pull-off, pull-out, stud-pull, etc. geometry. Nevertheless, this may not represent the real stresses that the electronics component, assembly or device faces in daily use. For example, the sample may experience shear stress rather than tensile ones. Therefore another fundamental adhesion test type is to prepare samples that experience shear stresses under load. Kivilahti et al. designed the so-called *grooved lap joint* setup that results in a realistic stress distribution within the test sample and, thus, the gained adhesion strengths from the test are also in line with the field results.

Not all the interfaces can be prepared in the tensile or shear test samples to provide fundamental quantitative results, thus engineers have to design new test geometries to study the performance of real interfaces to obtain at least qualitative information that enables them to rank interfaces. This ranking certainly can provide the basis for the evaluation of adequacy of pretreatments, coating methods and means to follow up the destructive effects of environmental stresses on the interfaces.

Coatings have been tested using peel, tape, torque, scratch and many other mechanical tests. None of these methods have yielded fundamental results although they may give some quantitative measure of the interfacial strength. Other than mechanical stressing may also be employed. Fluids–gases or liquids— have been used in blister testing in which it is been injected to the interface after a careful sample preparation process. Mechanical stresses may be produced in the sample by mechanical or acoustic vibrators and obviously by thermal cycling when thermally dissimilar materials are joined together. The adhesion of biomaterials and other very soft materials to surfaces is extremely difficult to measure without the risk of producing artefacts. For example, when soft silicone adhesion to LCP is tested by pushing it from one side, the modelling shows that the deformation is shifts significant when the indenter first deforms then penetrates into the silicone and eventually tears the silicone from the LCP substrate. This results in a mixed adhesion failure in which the edge is fractured from the bulk silicone and then the delamination takes place from the interface. For cell adhesion one may easily build a centrifugal test setup and adhesion of different cells on biomaterials can be determined by the rotational speed of the centrifuge.

References

1. G. Barnes, I. Gentle, *Interfacial Science* (Oxford University Press, Oxford, 2005)
2. E. Guggenheim, Thermodynamics of interfaces in systems of several components. Trans. Faraday Soc. **36**, 397 (1940)
3. P.K. Sharma, K. Hanumantha Rao, Analysis of different approaches for evaluation of surface energy of microbial cells by contact angle goniometry. Adv. Colloid Interface Sci. **98**, 341–463 (2002)
4. L.A. Girifalco, R.J.A. Good, Theory for the estimation of surface and interfacial energies. I. Derivation and application to interfacial tension. J. Phys. Chem. **61**, 904–909 (1957)
5. R.J. Good, L.A. Girifalco, G.A. Kraus, Theory for the estimation of surface and interfacial energies. II. Application to surface thermodynamics of teflon and graphite. J. Phys. Chem. **62**, 1418–1422 (1958)

6. R.J. Good, L.A.A. Girifalco, Theory for the estimation of surface, interfacial energies, III. Estimation of surface energies from contact anglecontact angle data. J. Phys. Chem. **64**, 561–565 (1960)
7. D. Li, A.W.A. Neumann, Reformulation of the equation of state for interfacial tensions. J. Colloid Interface Sci. **137**, 304–307 (1990)
8. J. K. Spelt, D. Li, in *Applied Surface Thermodynamics*, ed. by A. W. Neumann, J. K. Spelt, (Marcel Dekker, New York, 1996)
9. R.J. Good, J.A. Kvikstad, W.O. Balley, Anisotropic forces in a stretch-oriented polymer. J. Colloid Interface Sci. **35**, 314–327 (1971)
10. H.W. Fox, W.A.J. Zisman, The spreading of liquids on low energy surfaces. I. Polytetrafluoroethylene. J. Colloid Sci. **5**, 514–531 (1950)
11. C. J. van Oss, in *Polymer Surfaces and Interfaces*, vol. 2 ed by W. J. Feast, H. S. Munro, R. W. Richards (Wiley, Chichester, 1993)
12. S. Wu, *Polymer Interface and Adhesion* (Marcel Dekker, New York, 1982)
13. D.Y. Kwok, A.W. Neumann, Contact angle interpretation in terms of solid surface tension. Colloids Surf. A **161**, 31–48 (2000)
14. F.M. Fowkes, Additivity of intermolecular forces at interfaces. I. Determinationi of the contribution to surface and interfacial tensions of dispersion forces in various liquids. J. Phys. Chem. **67**, 2538–2543 (1963)
15. F. M. Fowkes, in *Acid–Base Interactions*, ed by K. L. Mittal, V. R. Anderson, Jr. (VPS, Utrecht, 1991) pp 93–115
16. D.K. Owens, R.C. Wendt, Estimation of surface free energy of polymers. J. Appl. Polym. Sci. **13**, 1741–1747 (1969)
17. D.H. Kaelble, K.C. Uy, Reinterpretation of organic liquid PTFE surface interactions. J. Adhes. **2**, 50–60 (1970)
18. C.J. van Oss, R.J. Good, M.K. Chaudhury, Additive and non-additive surface tension components and the interpretation of contact angle data. Langmuir **4**, 884–891 (1988)
19. C.J. van Oss, R.J. Good, M.K. Chaudhury, The role of van der Waals forces and hydrogen bonds in "hydrophobic interactions" between biopolymers and low energy surfaces. J. Colloid Interface Sci. **111**, 378–390 (1986)
20. S. Wu, Calculation of interfacial tension in polymer systems. J. Polym. Sci. Part C: Polym. Symp. **34**, 19–30 (1971)
21. C.J. van Oss, M.K. Chaudhury, R.J. Good, Interfacial lifshitz-van der waals and polar interactions in macroscopic systems. Chem. Rev. **88**, 927–941 (1988)
22. C.J. van Oss, L. Ju, M.K. Chaudhury, R.J. Good, Estimation of the polar parameters of the surface tension of liquids by contact angle measurement on gels. J. Colloid Interface Sci. **128**, 313–319 (1988)
23. C.J. van Oss, R.J. Good, Surface tension and the solubility of polymers and biopolymers: the role of polar and apolar interfacial free energies. J. Macromol. Sci. Chem. A **26**, 1183–1203 (1989)
24. C.J. van Oss, Mechanisms of conditions for repulsive van der Waals, repulsive hydrogen interactions. J. Dispers. Sci. Technol. **11**, 491–502 (1990)
25. C.J. van Oss, K. Arnold, R.J. Good, K. Gawrisch, S. Ohki, Interfacial tension, the osmotic pressure of solutions of polar polymers. J. Macromol. Sci. Chem. A **27**, 563–580 (1990)
26. C.J. van Oss, R.J. Good, H.J. Busscher, Estimation of the polar surface parameters of glycerol, formamide for use in contact angle measurements on polar solids. J. Dispers. Sci. Technol. **11**, 75–81 (1990)
27. C.J. van Oss, R.F. Giese Jr, R.J. Good, Re-evaluation of the surface tension components and parameters of polyacetylene from contact angle of liquids. Langmuir **6**, 1711–1713 (1990)
28. I.D. Morrison, Does the phase rule for capillary systems really justify an equation of state for interfacial tensions? Langmuir **7**, 1833–1836 (1991)
29. R.E. Johnson, R.H. Dettre, An evaluation of Neumann's "surface equation of state". Langmuir **5**, 293–295 (1989)

30. D. Mayers, *Surfaces, Interfaces and Colloids Principles and Applications* (VCH, New York, 1991), p. 433
31. C. van Oss, M. Chaudhury, R. Good, Monopolar surfaces. Adv. Colloid Interface Sci. **28**, 35 (1987)
32. C. van Oss, R. Good, M. Chaudhury, Mechanism of DNA (Southern) and protein (Western) blotting on cellulose nitrate and other membranes. J. Chomatog. **391**, 53 (1987)
33. G. Wolansky, A. Marmur, Apparent contact angle on rough surfaces: the Wenzel equation revisited. Colloid Surf. A **156**(1–3), 381 (1999)
34. G. Palasantzas, J.Th.M. de Hosson, Wetting of rough surfaces. Acta Mater. **49**, 3533 (2001)
35. Th. Uelzen, J. Müller, Wettability enhancement by rough surfaces generated by thin film technology. Thin Solid Films **434**(1–2), 311 (2003)
36. R.N. Wenzel, Resistance of solid surfaces to wetting by water. Ind. Eng. Chem. **28**, 988 (1936)
37. A.B.D. Cassie, S. Baxter, Wettability of porous surfaces. Trans. Farad Soc. **40**, 546 (1944)
38. J.N. Israelivich, M.L. Gee, Contact angles on chemically heterogeneous surfaces. Langmuir **5**, 288 (1989)
39. R. Hofer, M. Textor, N.D. Spencer, Imaging of surface heterogenieity by the microdroplet condensation technique. Langmuir **17**, 4123 (2001)
40. E. Chibowski, Surface free energy of a solid from contact angle hysteresis. Adv. Colloid Interface **103**(2), 149 (2003)
41. J.S. Kim, R.H. Friend, F. Cacialli, Surface energy, polarity of treated indium-tin-oxide anodes for polymer light-emitting diodes studied by contact-angle measurements. J. Appl. Phys. **86**(5), 2774 (1999)
42. C.W. Extrand, Water contact angle, hysteresis of polyamide surfaces. J. Colloid Interf. Sci. **248**(1), 136 (2002)
43. C.N.C. Lam, N. Kim, D. Hui, D.Y. Kwok, M.L. Hair, A.W. Neumann, The effect of liquid properties to contact angle hysteresis. Colloid Surf. A **189**(1–3), 265 (2001)
44. L. Weh, Self-organized structures at the surface of thin polymer films. Mater Sci. Eng. C **8–9**, 463 (1999)
45. X.F. Lu, J.N. Hay, Crystallization orientation and relaxation in uniaxially drawn poly(ethylene terephthalate). Polymer **42**(19), 8055 (2001)
46. S. Aida, S. Sakurai, S. Nomura, Strain-induced ordering of microdomain structures in polystyrene-block-polybutadiene-block-polystyrene triblock copolymers cross-linked in the disordered state. Polymer **43**(9), 2881 (2002)
47. L. Weh, Self-organized structures at the surface of thin polymer films. Mater Sci. Eng. C **8–9**, 463 (1999)
48. X.F. Lu, J.N. Hay, Crystallization orientation and relaxation in uniaxially drawn poly(ethylene terephthalate). Polymer **42**(19), 8055 (2001)
49. S. Aida, S. Sakurai, S. Nomura, Strain-induced ordering of microdomain structures in polystyrene-block-polybutadiene-block-polystyrene triblock copolymers cross-linked in the disordered state. Polymer **43**(9), 2881 (2002)
50. T. Cubaud, M. Ferminger, Advancing contact lines on chemically patterned surfaces. J. Colloid Interf. Sci. **269**, 171 (2004)
51. S. Breisch, B. de Heij, M. Löhr, M. Stelzle, Selective chemical surface modification of fluidic microsystems, characterization studies. J. Micromech. Microeng. **14**, 497 (2004)
52. C. Park, J. Yoon, E.L. Thomas, Enabling nanotechnology with self assembled block copolymer patterns. Polymer **44**(22), 6725 (2003)
53. P.G. de Gennes, Wetting: statics and dynamics. Rev. Mod. Phys. **57**(3), 827 (1985)
54. F.G. Yost, P.A. Sackinger, E.J. O'Toole, Energetics and kinetics of dissolutive wetting processes, Acta Mater **46**(7),1998
55. S. Kalogeropoulou, C. Rado, N. Eustathopoulos, Mechanisms of reactive wetting: The wetting to non-wetting case. Scripta Mater **41**(7), 723 (1999)
56. W.J. Boettinger, C.A. Handwerker, U.R. Kattner, in *The mechanics of solder alloy wetting and spreading*, ed. by F.G. Yost, F.M. Hosking, D.R. Frear. Reactive wetting and intermetallic formation, (Van Nostrand Reinhold, New York, 1993)

57. G.W. Gastellan, *Physical chemistry*, 2nd edn. (Addison-Wesley Publishing Company, Massachusetts, 1971)
58. R. Defay, I. Prigogine, A. Bellemans, D.H. Everett, *Surface Tension and Adsorption, Longmans* (Green & Co LTD, London, 1966)
59. K.K. Wang, Underfill of flip chip on organic substrate: viscosity, surface tension, and contact angle, J. Microelectron. Reliab. **42**, 293 (2002)
60. A.M. Alteraifi, D. Sherif, A. Moet, Interfacial effects in the spreading kinetics of liquid droplets on solid substrates. J. Colloid Interf. Sci. **264**(1), 221 (2003)
61. L.H. Tanner, The spreading of silicone oil drops on horizontal surfaces. J. Phys. D Appl. Phys. **12**(9), 1473 (1975)
62. A. Seaver, J. Berg, Spreading of a droplet on a solid surface. J. Appl. Polym. Sci. **52**, 431 (1994)
63. D.T. Wasan, A.D. Nikolov, H. Brenner, Droplets speeding on surfaces. Science **291**, 605 (2001)
64. M.P.K. Turunen, T. Laurila, J. Kivilahti, Evaluation of the surface free energy of spin-coated photodefinable epoxy. J. Polym. Sci: Part B. Polym. Phys. **40**, 2137 (2002)
65. J. Ge, R. Tuominen, J. Kivilahti, Adhesion of electrolessly-deposited copper to photosensitive epoxy. J. Adhes. Sci. Technol. **15**(10), 1133 (2001)
66. J. Ge, M.P.K. Turunen, J. Kivilahti, Surface modification of a liquid crystalline polymer for copper metallization. J. Polym. Sci.: Part B. Polym. Phys. **41**, 623 (2003)
67. J. Ge, M.P.K. Turunen, J. Kivilahti, Surface modification and characterisation of photodefinable epoxy/copper systems. Thin Solid Films **440**(1–2), 198 (2003)
68. J. Ge, M.P.K. Turunen, M. Kusevic, J. Kivilahti, Effects of surface treatments on adhesion of copper to a hybrid polymer material. J. Mater Res. **18**, 2697 (2003)
69. M.P.K. Turunen, in *Interfacial compatibillity of polymer-ased structures in electronics*, Doctoral Thesis, Helsinki University of Technology, 2004, p. 62
70. D. Schröer, R.J. Nichols, H. Meyer, Pretreatment of polymer surfaces : the crucial step prior to metal deposition. Electrochim. Acta **40**, 1487 (1995)
71. X. Roizard, M. Wery, J. Kirmann, Effects of alkaline etching on the surface roughness of a fibre-reinforced epoxy composite. Compos. Struct. **56**, 223 (2002)
72. K.-W. Lee, S.P. Kowalczyk, J.M. Shaw, Surface modification of PMDA-oxydianiline polyimide. Surface-adhesion relationship. Macromolecules **23**, 2097 (1990)
73. R.R. Thomas, S.L. Buchwalter, L.P. Buchwalter, T.H. Chao, Organic chemistry on a polyimide surface. Macromolecules **25**, 4559 (1992)
74. Z. Wang, A. Furuya, K. Yasuda, H. Ikeda, T. Baba, M. Hagiwara et al., Adhesion improvement of electroless copper to a polyimide film substrate by combining surface microroughening and imide ring cleavage. J. Adhes. Sci. Technol. **16**, 1027 (2002)
75. I. Ghosh, J. Konar, A.K. Bhownick, Surface properties of chemically modified polyimide films. J. Adhes. Sci. Technol. **11**, 877 (1997)
76. P.K. Chu, J.Y. Chen, L.P. Wang, N. Huang, Plasma-surface modification of biomaterials. Mater Sci. Eng. R **36**, 143 (2002)
77. K. Richter, M. Orfert, K. Drescher, Anisotropic patterning of copper-llaminated polyimide foils by plasma etching. Surf. Coat. Tech. **97**, 481 (1997)
78. C.E. Park, B.J. Han, H.E. Bair, Humidity effects on adhesion strength between solder ball and epoxy underfills. Polymer **38**(15), 3811 (1997)
79. M.-L. Sham, J.-K. Kim, Adhesion characteristics of underfill resins with flip chip package components. J. Adhes. Sci. Technol. **17**, 1923 (2003)
80. K.L. Mittal, H.R. Andersson Jr., *Acid-Base Interactions* (VPS, Utrecht, 1991)
81. J.M. Burkstrand, Metal-polymer interfaces: adhesion and X-ray photoemission studies. J. Appl. Phys. **52**, 4795 (1981)
82. J.F. Friedrich, W.E.S. Unger, A. Lippitz, I. Koprinarov, G. Kuhn, St. Weidner, L. Vogel, Chemical reactions at polymer surfaces interacting with a gas plasma or with metal atoms—their relevance to adhesion. Surf. Coat. Technol. **116/119**, 772–786 (1999)

83. L.J. Martin, C.P. Wong, Chemical and mechanical adhesion mechanisms of sputter - deposited metal on epoxy dielectric for high density interconnect printed circuit boards. IEEE Trans. Comp. Packag Technol. **24**, 416 (2001)
84. Y. Jugnet, J.L. Droulas, D.T. Minh, A. Pouchelon, in: E. Sacher, J.J. Pireaux, S.P. Kowalczyk (eds.), Metallization of Polymers, ACS Symposium Series 440, 1990, pp. 467–484
85. A.J. Kinloch, The science of adhesion. J. Mater Sci. **17**(3), 617–651 (1982)
86. D.R. Lefebre, K.M. Takahashi, A.J. Muller, V.R. Raju, Degradation of epoxy coatings in humid environments: the critical relative humidity for adhesion loss. J. Adhes. Sci. Technol. **5**(3), 201 (1991)
87. O. Negele, W. Funke, Internal stress and wet adhesion of organic coatings. Prog. Org. Coat. **28**, 285–289 (1996)
88. A.J. Kinloch, M.S.G. Little, J.F. Watts, The role of the interphase in the environmental failure of adhesive joints. Acta Mater **48**, 4543–4553 (2000)
89. E. Breslauer, T. Troczynski, Determination of the energy dissipated during peel testing. Mat. Sci. Eng. A **302**, 168 (2001)
90. K.L. Mittal (ed.), *Adhesion Measurement of Films and Coatings* (VPS, Utrecht, 1995)
91. A.J. Kinloch, *Adhesion and Adhesives: Science and Technology* (Chapman and Hall, London, 1987)
92. A.J. Kinloch, The science of adhesion. J. Mater Sci. **17**, 617 (1982)
93. J. Valli, A review of adhesion adhesion test methods for thin hard coatings. J. Vac. Sci. Technol. A **4**(6), 3007 (1986)
94. M.S. Kafkalidis, M.D. Thouless, The effects of geometry and material properties on the fracture of single lap-shear joints. Int. J. Solids Struct. **39**, 4367 (2002)
95. U. Rekners, M. Kalnins, Evaluation of the protective properties of organic coatings by using tape and blistering tests. Prog. Org. Coat. **38**, 35–42 (2000)
96. H. Ollendorf, D. Schneider, A comparative study of adhesion test methods for hard coatings. Surf. Coat. Tech. **113**, 86 (1999)
97. A.J. Kinloch (ed.), *Durability of Structural Adhesives* (Elsevier Applied Science, London, 1983)
98. M.P.K. Turunen, P. Marjamäki, M. Paajanen, J. Lahtinen, J. Kivilahti, in Pull-off test in the assessment of adhesion at printed wiring board metallisation/epoxy interface , Microelectr Reliab **44**, 993, (2004)
99. A. Miszczyk, T. Schauer, Electrochemical approach to evaluate the interlayer adhesion of organic coatings. Prog. Org. Coat. **52**, 298–305 (2005)
100. T. Nguyen, E. Byrd, D. Bentz, C. Lin, In situ measurement of water at the organic coating/substrate interface. Prog. Org. Coat. **27**, 181–193 (1996)
101. M. Stratmann, A. Leng, W. Furbeth, H. Streckel, H. Gehmecker, K.-H. Grosse-Brinkhaus, The scanning Kelvin probe; a new technique for the in situ analysis of the delamination of organic coatings. Prog. Org. Coat. **27**, 261–267 (1996)
102. J.N. Murray, Electrochemical test methods for evaluating organic coatings on metals: an update. Part III: Multiple test parameter measurements. Prog. Org. Coat. **31**, 375–391 (1997)
103. R.A. Gledhill, A.J. Kinloch, S.J. Shaw, Effect of relative humidity on the wettability of steel surfaces. J. Adhes. **9**(1), 81 (1977)
104. S. Niiranen, in *Deterioration of adhesion between plasma treated LCPLCP and silicone under accelerated environmental stress tests*, Master's Thesis, Aalto University, in Finnish, 2011, 63 pp
105. D. Tabor, Basic principles of adhesion. Rep. Prog. Appl. Chem. **36**, 621 (1951)
106. D.J. Alner (ed.), *Aspects of Adhesion*, vol. 5 (University of London Press, London, 1969)
107. A.A. Griffith, Phenomena of rupture and flow in solids. Phil. Trans. **221A**, 163 (1920)
108. E.H. Andrews, A.J. Kinloch, Mech. adhes. fail. II. Proc. Roy. Soc. **A332**, 385 (1973)
109. D.E. Packham, Work of adhesion: contact angles and contact mechanics. Int. J. Adhes. **16**(2), 121–128 (1996)

110. A. Ahagon, A.N. Gent, Effect of interfacial bonding on the strength of adhesion. J. Polym. Sci. Polym. Phys. **13**, 1285 (1975)
111. D.E. Peckham, J.R.G. Evans, P.R. Davies, The effect of temperature of test on the adhesion of polyethylene coatings applied to metals as a hot melt. J. Adhes. **13**, 29 (1981)

Chapter 6
Evolution of Different Types of Interfacial Structures

In this chapter several cases where the thermodynamic-kinetic approach introduced in Chap. 4 and interfacial adhesion studies introduced in Chap. 5 are utilised to study different cases dealing with interfacial reactions will be presented. The examples show how powerful the thermodynamic–kinetic method is in rationalising the formation of reaction layer sequences, effect of impurities or alloying elements on the reactions and so on. Since the approach is completely general, the only limitation in its application is the lack of relevant thermodynamic and kinetic data. Nevertheless, even without proper data the method can be still used to provide general guidelines for reasoning. This is seen in some of the cases where the proper data is not available, but the method still provides considerable insight to the problem at hand.

6.1 Examples of Metal–Metal Interfaces

Metals are widely used in many technological fields due to their favourable physical and chemical properties. They form interfaces with other metals as well as with other material groups, such as ceramics and polymers. In this section different metal/metal reaction couples (solid/solid as well as solid/liquid) are discussed. In the subsequent chapters interfaces between metals and other classes of materials (polymers and ceramics) are discussed.

6.1.1 Au/Al in Wire Bonding

Reactions between Au wire and Al pad metallisation have been studied widely and the degradation of the bonds in high temperature ageing between 150–200°C is reported in many studies [1–4]. A few different models for failure mechanisms of

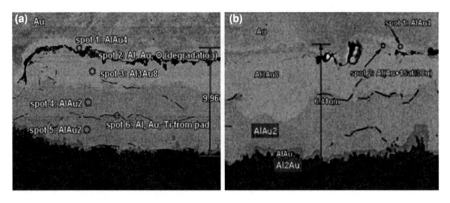

Fig. 6.1 Micrograph of the Au–Al bond after 1,000 h at 150°C for (**a**) 99.99% Au wire (**b**) 99% Au wire with 1 wt% Cu

the bonds are proposed in these studies. It has been suggested that Kirkendall voids due to unequal material fluxes have contributed to the failure [2]. Other communications suggest that contamination-induced corrosion plays a major role in reducing the bond strength [4]. Specifically, the Au$_4$Al intermetallic has been reported to be susceptible to dry corrosion, whereas other AuAl intermetallics react either more slowly or are not affected [3]. Owing to this high temperature degradation, the general opinion has been that an Au wire bonding is a viable option only up to 150°C junction temperature applications. Especially, reliability problems associated with polymer packages have become a major concern. The addition of a small amount of Cu in the wire material has been shown to have a positive effect on the wire bond strength, as measured after thermal ageing [5, 6]. However, it is not actually clear what the exact role of Cu in the reactions is.

Figure 6.1 presents a typical Au–Al bond failure with 4 N (99.99% Au) Au wire after annealing for 1,000 h at 150°C and the effect of Cu addition. Initial pad metallisation is fully consumed, and a typical IMC layer sequence of Au/Au$_4$Al/Au$_8$Al$_3$/Au$_2$Al has been formed. The bond is fractured through the entire interface along the Au-rich Au$_4$Al intermetallic. With the addition of 1 wt% Cu in the wire, a similar (but not identical—see below) phase sequence is found after aging, but there is no continuous fracture within the intermetallics. The main difference in the compositional analysis is the presence of Cu in the Au$_4$Al layer. Within Au$_4$Al IMC (or next to it) a distinctive layer with a darker contrast can be seen, which was found to contain about 11 at % Cu.

The impact of Cu on the microstructural evolution and degradation of Au–Al bonds has been studied mainly experimentally so far. Gam et al. [5] observed a formation of a Cu-containing layer between Au$_4$Al and Au, and suggested that this would decrease the Au–Al reaction rate. Chang et al. [6] reported two Cu-containing phase compositions to be formed next to Au; Au$_4$(Al,Cu) and (Au,Cu)$_4$Al. With reliable thermodynamic description of the system and diffusion data it would be possible to clarify the formation of IMC phases and the impact of

6.1 Examples of Metal–Metal Interfaces

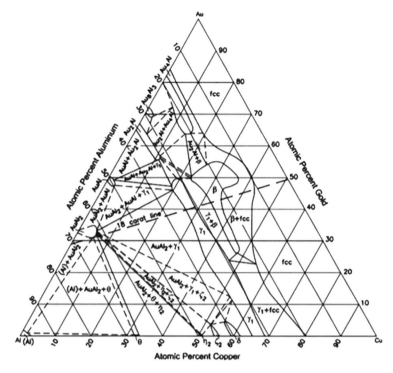

Fig. 6.2 Experimentally determined isothermal section at 500°C from the Al–Au–Cu system [7]

Cu on their growth and properties. Unfortunately, the data available is very limited. One of the very few ternary phase diagrams for the Au–Al–Cu system known to the authors is one which has been experimentally determined at 500°C by Levey et al. [7], and is shown in Fig 6.2.

Despite the high temperature at which this phase diagram was completed, some general conclusions can be made. As shown in Fig. 6.2, all AuAl intermetallics dissolve some Cu, but the highest dissolution of Cu is found in the Au$_4$Al phase. At 500°C, up to 15 at.% Cu is dissolved in the Au$_4$Al phase. It is expected that the trend of the dissolution remain at lower temperatures, although the maximum solubility is most likely lower. Therefore it is anticipated that Cu can stabilise the Au$_4$Al intermetallic strongly. It is to be noted that the ternary solubility shown in the diagram (Fig. 6.2) for Au$_4$Al indicates that Cu enters the Au-sublattice [(Au,Cu)$_4$Al, as the Al content in the Au$_4$Al remains constant. This contradicts the results obtained by Chang et al. [6], who reported also the observation of the Au$_4$(Al,Cu) compound. As Au and Cu are in the same group in the periodic table and chemically very similar, the substitution of Au with Cu in the Au-sublattice of Au–Al IMCs seems highly probable.

According to the ternary isotherm at 500°C, the layer sequence in Figs. 6.1a and b, where the local equilibrium between Au$_2$Al, Au$_8$Al$_3$ and Au$_4$Al exists, is

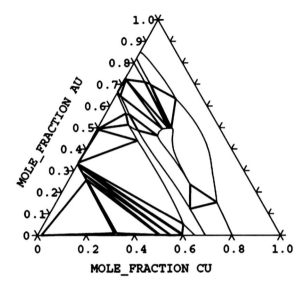

Fig. 6.3 Calculated isothermal section of the Al–Au–Cu system at 500°C

possible only if Cu solubility to Au$_4$Al is restricted to about 8 at.% or less. This is because if the amount of Cu is higher than that, the local equilibrium between Au$_4$Al and Au$_8$Al$_3$ is no longer possible and Au$_4$Al comes directly into local equilibrium with Au$_2$Al. According to this reasoning the experimentally observed structure would not be an equilibrium one, since about 11 at.% of Cu was detected inside Au$_4$Al. The temperature is, of course, considerable higher in the diagram than in the case shown in Fig. 6.1 and only tentative conclusions based on the isothermal section can be made. On the other hand, a high amount of Cu experimentally found within the Au–Al IMC region indicates that there is significant Cu solubility to Au–Al IMC's also at lower temperatures.

We have made preliminary calculations of the Al–Au–Cu system at several different temperatures including 500 and 150°C (Figs. 6.3 and 6.4). Comparison between our calculated diagram and that experimentally determined at 500°C in [7] reveals that there are slight differences in equilibria between these two isothermal sections. The foremost different are in the local equilibria conditions between Au$_4$Al, Au$_8$Al$_3$ and Au$_2$Al. First, as can be seen from Fig. 6.3 there is no direct equilibrium between Au$_4$Al and Au$_2$Al, not even when about 20 at.% of Cu is dissolved into Au$_4$Al. Second, it is to be noted that the solubility of Cu to Au$_8$Al$_3$ seems to be much less in Fig. 6.3 than that shown in Fig. 6.2. Thus, the local equilibrium between Au$_4$Al and Au$_8$Al$_3$ (Fig. 6.1) is possible even with large amounts of Cu inside Au$_4$Al. The isothermal section at 150°C is shown in Fig. 6.4. As can be seen from the diagram the high solubility of Cu to Au$_4$Al seems to be retained at lower temperatures also. The estimated value is about 10–12 at.% of Cu in Au$_4$Al at 150°C. Based on the diagram the local equilibrium between Au$_4$Al with about 10 at.% dissolved Cu and Au$_8$Al$_3$ (Fig. 6.1) is possible at 150°C. However, if the amount of dissolved Cu would be higher than 10 at.%, ternary β

6.1 Examples of Metal–Metal Interfaces

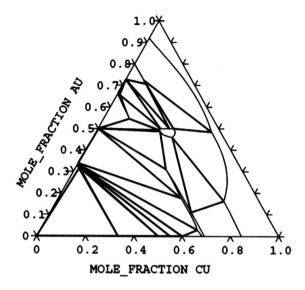

Fig. 6.4 Calculated isothermal section from the Al–Au–Cu ternary phase diagram at 150°C

should be involved in the local equilibrium. This is not experimentally observed (Fig. 6.1).

Based on the extensive solubility of Cu in Au$_4$Al one can expect that Cu stabilises Au$_4$Al substantially and accordingly the dry corrosion of Au$_4$Al will be reduced by the increased stability of the phase. One can divide the oxidation of Au$_4$Al roughly into two categories: ball bonds embedded in moulding compounds and bare wire bonds (exposed directly to environment). When Au–Al wire bonded devices are embedded in molding compound, the presence of halides in polymers has been observed to assist degradation of the wire bonds by oxidising the Au$_4$Al IMC [4]. Thus, if the reaction

$$Au_4Al + X \text{ (such as Br, Cl, F)} AlX_3 + 4 \text{ Au}$$

and the subsequent oxidation of the AlX$_3$ to Al$_2$O$_3$ (and liberation of reactive X species to continue the corrosion reaction) would eventually slow down or even completely cease to take place due to the increased stability of the (Au,Cu)$_4$Al phase in the presence of Cu, the reliability of the wire bonds should improve. Oxidation of Au$_4$Al in bare devices has been identified to cause bond lift failures [3, 8, 9]. Because of the increased stability of the (Au,Cu)$_4$Al, its oxidation in bare ball bonds (not embedded in any molding compound) should also slow down and therefore increased reliability of devices can be expected in such cases as well. Indeed, as from Fig. 6.1b can be seen, with Cu addition the oxidation of Au$_4$Al seems to be hindered as the spot analyses do not reveal any significant traces of oxygen.

On the other hand, in the absence of the suitable phase diagram and reliable activity data under 200°C, possible diffusion paths across the Au(Cu)—Au$_8$Al$_3$ cannot be evaluated, and therefore we cannot exclude the possibility that the effect of Cu on the IMC growth is related to the reduction of Au diffusion through the reaction layer sequence. When comparing Figs. 6.1a and b it can be noted that the total IMC layer thickness is less in the case of Cu alloyed wire than with the "pure" Au wire. As Au atoms must diffuse through $(Au,Cu)_4Al$ to reach other Au–Al IMCs, any changes in the diffusion flux inside $(Au,Cu)_4Al$ will have an effect on the growth of the other IMC layers. Furthermore, from Fig. 6.1b one can see the formation of AuAl and $AuAl_2$ intermetallics, whereas in Fig. 6.1a they are absent. The layer sequence in Fig. 6.1b is typically associated with thin film couples with excess Al [10]. Thus, this can be taken as an indication that Au diffusion is somewhat retarded when Au wire is alloyed with Cu, since in this case there is excess Au (typically associated with phase sequence seen in Fig. 6.1a). Based on the calculated preliminary activity data in Al–Au–Cu system, it seems that the presence of Cu in Au_4Al lowers the activity of Au and subsequently would reduce the driving force for Au diffusion through the next more Al-rich Au–Al IMC layers. Another probable reason for the influence of Cu on the diffusion of Au in $(Au,Cu)_4Al$ might be that the presence of Cu influences the defect population of the IMC in question. It is well established that in ordered compounds minor changes in defect population will have major effects on the diffusion kinetics [11, 12]. In highly ordered alloys or compounds random motion of vacancies is not possible, as it would disrupt the equilibrium ordered arrangement of atoms on lattice sites and would be energetically unfavourable. Thus, all changes in atomic environment are expected to strongly influence the diffusion of elements in the intermediate compound in question.

The generation of more thermodynamic and diffusion kinetic information for the Au–Al–Cu system is necessary to fully clarify the reactions and the failure mechanisms. This is specifically important since addition of small amounts of Cu or other additives may be the key to increase the high temperature reliability of Au wire bonds. The thermodynamic-kinetic work is currently carried out in our group.

6.1.2 Sn-Based Solder Versus UBM in Flip Chip

Despite the introduction of Cu metallised ICs during the 1990s, Al metallised ICs are still produced in high volumes. When these are used, for example, in flip chip (FC) applications, additional metal layers must be used to ensure solderability. These metallic layers form the so-called Under Bump Metallurgy (UBM). This metallurgy consists of different thin metal layers needed in attaching bumps onto chip with Al-metallisations. Under Bump Metallurgy protects also the remaining unprotected areas, for instance, contact pads of the chip from corrosion during use of the device. Appropriate UBM is essential in order to achieve reliable flip chip connections. Several different UBM structures have been developed. Typically there is at least an adhesion layer (TiW or similar) next to the Al contact pad, then a solderability layer (Cu, Ni, Ni(V), Ni(P) or similar) and an protection layer against oxidation (Au or similar). It is to be noted that also more complicated structures are possible. Nevertheless, the

6.1 Examples of Metal–Metal Interfaces 141

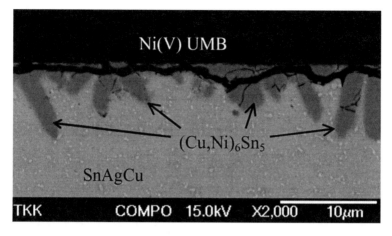

Fig. 6.5 Cracking of the (Cu,Ni)$_6$Sn$_5$ reaction layer during drop testing

compatibility of the UBM and the solder as well as printed wiring board (PWB) metallisation used is of great importance when the reliability of FC interconnections is considered. In this chapter a few examples of different problems associated with UBM and their rationalisation with the thermodynamic–kinetic method are presented.

6.1.2.1 Case 1: Effect of Cu Concentration on the First IMC Formed on Top of Ni Metallisation

When using liquid non-Cu-containing solders, such as SnPb, SnAg, or SnSb the Ni$_3$Sn$_4$ phase is typically the first phase to form in the reaction between Ni and Sn. However, when using Pb-free solders that include small amounts of Cu the situation changes so that the first phase to form is (Cu,Ni)$_6$Sn$_5$. This situation can occur also when for example Cu metallised contact pads are used in the PWB-side, even though solder itself does not contain Cu. This Ni containing Cu$_6$Sn$_5$ intermetallic compound (IMC) is very brittle and has been observed to cause severe reliability problems, especially under mechanical shock loading (see Fig. 6.5). On the other hand, if solders do not contain enough Cu to enable the formation of (Cu,Ni)$_6$Sn$_5$, the (Ni,Cu)$_3$Sn$_4$ forms on top of Ni metallisation. Therefore, it is of great importance to determine the critical Cu concentration, which causes the formation of (Cu,Ni)$_6$Sn$_5$ and its dependence on temperature and the presence of other alloying elements.

The thermodynamic-kinetic approach can be used for predicting the intermetallic reaction products in Cu-containing lead-free solders with Ni metallisation. Figure 6.6 shows the Sn-rich corner of the isothermal section of the Sn–Cu–Ni phase diagram calculated at 250°C. If the Cu content of the solder is, for example, 0.3 wt% (indicated with A in Fig. 6.6) (Ni,Cu)$_3$Sn$_4$ nucleates at the Ni|solder interface because the dissolution of Ni into the solder changes the local

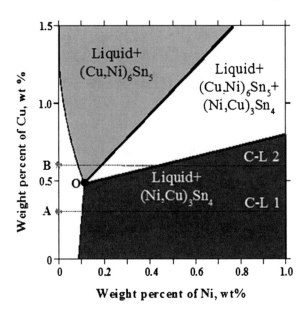

Fig. 6.6 Tin-rich corner of the Sn–Cu–Ni phase diagram at 250°C

nominal composition towards Ni along to the contact line 1 (C-L 1). Hence, in the cases when the local nominal composition (i.e. the average composition of solder near the interface) enters the dark grey area the liquid solder comes in local equilibrium with (Ni,Cu)$_3$Sn$_4$. On the other hand, if the original Cu content of the solder is higher, say 0.6 wt% (indicated with B in Fig. 6.6), the dissolution of Ni (C-L 2) leads to local equilibrium between liquid and (Cu,Ni)$_6$Sn$_5$ (the light gray area in Fig. 6.6). In addition, as the dissolution of Ni can continue until the metastable solubility, which is typically 2–3 times the stable solubility (introduced in Chap. 4) is reached, it is also possible that the local nominal composition enters the three-phase triangle and then (Cu,Ni)$_6$Sn$_5$, Ni,Cu)$_3$Sn$_4$ and the liquid solder (with the composition marked with O) can come in local equilibrium.

However, when the temperature is not constant (as in soldering) the phase equilibria will change as a function of temperature. These changes can be examined with the help of Fig. 6.7, in which the apex of the three-phase triangle (point O in Fig. 6.6), i.e. the critical Cu content of the solder, is presented as a function of temperature. Above (to the left of) the dotted line the primary intermetallic is (Cu,Ni)$_6$Sn$_5$ and below (to the right of) it the primary intermetallic is (Ni,Cu)$_3$Sn$_4$. The circles [(Cu,Ni)$_6$Sn$_5$] and square [(Ni,Cu)$_3$Sn$_4$] represent the experimental results shown in Fig. 6.8.

For example, when the Cu content of the solder is 0.5 wt% and reflow peak temperature is 240°C (Cu,Ni)$_6$Sn$_5$ will form first, but when the temperature is increased up to 270°C (Ni,Cu)$_3$Sn$_4$ comes in local equilibrium with the solder, as seen from Fig. 6.7. and experimentally verified as shown in Figs. 6.8a and b respectively. The same change is also observed (see Fig 6.8c) if the temperature is kept at 240°C, but the Cu content of the solder is decreased to 0.3 wt%. When

Fig. 6.7 Calculated critical Cu content in liquid Sn to change interfacial reaction product from (Ni,Cu)$_3$Sn$_4$ to (Cu,Ni)$_6$Sn$_5$, as a function of temperature together with the experimental points

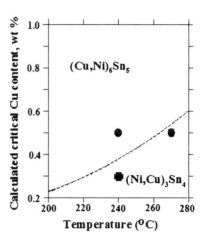

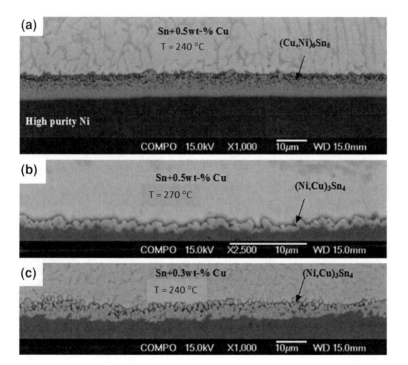

Fig. 6.8 a Sn + 0.5wt-% Cu/Ni diffusion couple annealed at 240°C, **b** Sn + 0.5wt-%Cu/Ni diffusion couple annealed at 270°C, and **c** Sn + 0.3wt-%Cu/Ni diffusion couple annealed at 240°C for 60 min

comparing Figs. 6.8a and b, it is interesting to note that the IMC thickness is clearly reduced when it is changed from (Cu,Ni)$_6$Sn$_5$ to (Ni,Cu)$_3$Sn$_4$, even when the temperature is increased. Therefore it can be concluded that the effect

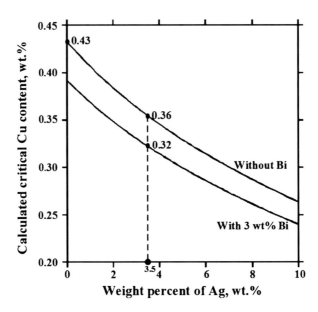

Fig. 6.9 Effect of alloying Ag and Bi to Sn on the critical Cu content at T = 250°C

of temperature on the critical Cu composition is significant and thus it is possible to explain the $(Cu,Ni)_6Sn_5$ "precipitates" on the top of the $(Ni,Cu)_3Sn_4$ observed in [13–16] when the Cu content of the SnAgCu solder was 0.4 wt%.

The addition of other alloying elements will change the critical Cu content for $(Cu,Ni)_6Sn_5$ formation. Ho et al. [14] suggested that the existence of Ag in Sn–Ag–Cu solder does not have any effect, since it dissolves in neither $(Ni,Cu)_3Sn_4$ nor $(Cu,Ni)_6Sn_5$. This is not completely true, because the activities of all elements, not only in the compounds themselves but also in a liquid in (local) equilibrium with the compounds, determine the relative stabilities of $(Ni,Cu)_3Sn_4$ and $(Cu,Ni)_6Sn_5$. In order to investigate this issue in more detail, we have collected the thermodynamic interaction parameters in the liquid for the Sn–Cu–Ni–Ag system available in the literature and calculated the critical Cu content as a function of Ag content of the solder. As shown in Fig. 6.9, the critical Cu content decreases when the Ag content of the solder is increased. Bismuth is another much-used alloying element in lead-free solders. If we add, say, 3 wt% Bi to the liquid alloy the critical composition of the liquid is further decreased. It means that both Ag and Bi will reduce the critical Cu content even though they do not dissolve in either of the compounds. An interesting subject for future studies is the effect of other alloying elements on the activities of components in both liquid and intermetallic compounds.

6.1.2.2 Case 2 Effects of Ni on the Growth and Properties of IMC Layers in Cu–Sn System

Despite the excellent solderability of Cu with wide range of different solders, Ni is used as a diffusion barrier on top of Cu in many cases to slow down the reaction kinetics between the UBM and the Sn-based solders. If the good wetting properties of Cu could be feasibly combined with the slower dissolution and reaction kinetics of Ni, Cu(Ni) alloys could be used as "new" metallisation materials. In relation to this issue, Ohriner [17] has carried out an extensive investigation about reactions between various Cu-based alloys and Sn containing solders at temperatures between 150 and 250°C. The substrates were exposed to liquid solder for 5 s (at temperature 30°C above the liquidus temperature of the solder) and subsequently annealed in solid-state at different temperatures 150–225°C [17]. He observed an interesting behaviour, when Ni-containing alloys of Cu were soldered with Sn5Ag, Sn5Sb and SnPb$_{eut}$ solders. The rate of intermetallic formation was found to be dependent on Ni concentration and exhibited the maximum value in the range of 5–10 at.% Ni. The details of the compositional dependence varied with solder composition and exposure temperature, but the maximum rate of the intermetallic growth was always around 5–10 at.% of Ni in Cu [17]. The change in the reaction layer thickness was suggested to be related to changes in the composition of the intermetallic compound. Ohriner proposed that the addition of nickel to the η-phase might have important effects on the concentration of structural vacancies present in the intermetallic compound [17]. It should also be noted that Ohriner observed the nonuniformity of the IMC layer to increase with increasing annealing times [17]. Another important observation made by Ohriner was that the only IMC to grow between Ni-containing alloys of Cu and 95Sn-5Ag, 95Sn-5Sb and Sn–Pb solders was $(Cu,Ni)_6Sn_5$ within the resolution limits of SEM [17]. Thus, the addition of Ni to Cu suppressed the growth of Cu_3Sn almost completely. He suggested that the absence of Cu_3Sn was the result of either the thermodynamic or kinetics effects of Ni [17].

An interesting observation has been made recently by Paul [18] when studying solid-state reactions (T = 225°C) between pure Sn and Cu conductor metal alloyed with 5–15 at.% Ni. He also detected significant increase in the growth rate of $(Cu,Ni)_6Sn_5$ within this concentration range, as well as the absence of Cu_3Sn. Interestingly, when he etched the Sn away to reveal the $(Cu,Ni)_6Sn_5$, he observed that when Cu was alloyed with 5–15 at.% of Ni, the grain size of the $(Cu,Ni)_6Sn_5$ was more than one order of magnitude smaller than in the case of pure Cu [18]. This could explain the observed growth behaviour, since the faster grain boundary diffusion would result in a higher growth rate in solid-state. This is easily understood via the concept of effective diffusion coefficient introduced in Sect. 4.3.1. Moreover, high diffusion flux in $(Cu,Ni)_6Sn_5$ can suppress the growth of Cu_3Sn by making the Ni-containing η-phase kinetically very stable. We have also found the same behaviour (i.e. high growth rates of $(Cu,Ni)_6Sn_5$) within the same concentration range as described above in solid-state as well as in solid/liquid reaction couples.

Ni has also been used as an alloying element in solders. Amagai [19] added 0.01–0.03 wt% Ni to Sn3Ag (in wt%) and let it react with a Cu substrate in liquid/solid reaction couple. He observed that the grain structure of $(Cu,Ni)_6Sn_5$ after one reflow was refined and did not coarsen even after four reflows (in comparison to Sn3Ag/Cu reaction couple without Ni). Unfortunately, no solid-state annealing was made to investigate the growth behaviour in solid-state, i.e. to see how the small grain size of $(Cu,Ni)_6Sn_5$ would affect the growth rate of the IMC layer. Gao et al. [20] investigated the effect of addition of 0.1 wt% of Ni to Sn3.5Ag (wt%) solders on the interfacial reactions with a Cu substrate. Addition of Ni resulted in the formation of $(Cu,Ni)_6Sn_5$ and an elevated growth rate of the interfacial IMC, both in liquid and solid-state. Also the Cu_3Sn layer thickness was reduced during the solid-state annealing with respect to the reference Sn-Ag/Cu couple. They also observed that the alloying with Ni reduced the grain size of $(Cu,Ni)_6Sn_5$ markedly, a result consistent with other results from the literature [17, 18]. Thus, again the contribution from the grain boundary diffusion might explain the observed increase in the growth rate of $(Cu,Ni)_6Sn_5$.

Xu et al. [21] investigated the elastic modulus and hardness of the Cu_6Sn_5, Ni_3Sn_4, $(Cu,Ni)_6Sn_5$ and $(Ni,Cu)_3Sn_4$ after reflow and after annealing at 125°C by the nanoindentation continuous stiffness measurement (CSM) technique. The experiments were carried out by using 95.5%Sn3.8%Ag0.7%Cu (wt%) solder and Cu or Ni(P)/Au substrates. As it is known, the Cu content in the solder used in the investigation is more than enough to induce the formation of a $(Cu,Ni)_6Sn_5$ layer on Ni (or Ni(P)) substrate (see above). Thus, there were large amounts of Ni available to be incorporated into the $(Cu,Ni)_6Sn_5$ compound layer. Unfortunately, chemical analysis to determine the Ni content in the $(Cu,Ni)_6Sn_5$ layer was not done in [21]. They found out that the elastic modulus of Ni-containing $(Cu,Ni)_6Sn_5$ was about two times that of Cu_6Sn_5 and during solid-state annealing (500 h), it was reduced to about 70% of the value after reflow (from 207 to 146 GPa) [21]. The hardness of the Ni-containing $(Cu,Ni)_6Sn_5$ was also higher than that of Cu_6Sn_5 and it also dropped after solid-state annealing again to about 70% of the value after reflow (from 10.07 to 7.31 GPa) [21]. Thus, there is also a significant contribution to the mechanical properties of the Cu_6Sn_5 IMC when Ni is added. Basically, the presence of Ni makes the Cu_6Sn_5 IMC stronger and harder. This is good to keep in mind when results from Ni-alloying experiments are analysed.

6.1.2.3 Thermodynamic–Kinetic Analysis of the Effect of Ni on the Solid-State IMC Growth in Cu–Sn System

Based on results from the literature (as well as our own experimental and theoretical investigations) the following statements concerning the influence of Ni on the Cu–Sn IMC growth in solid-state can be made. (i) The addition of small amounts (0.1 at.% of Ni) to Cu conductor or to solder decreases the total thickness of the IMC layer to about half of that in the Cu/Sn diffusion couple and decreases markedly the ratio of Cu_3Sn–Cu_6Sn_5. If more than about 0.4 at.% of Ni is added to

6.1 Examples of Metal–Metal Interfaces

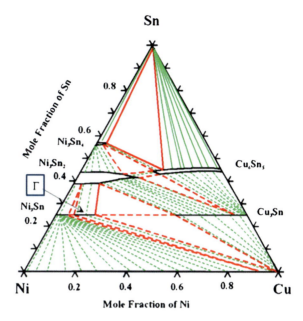

Fig. 6.10 Isothermal section at 125°C from the assessed Sn–Cu–Ni ternary phase diagram with the superimposed diffusion path

liquid Sn-based solder, Ni_3Sn_4 precipitates out of the solder matrix, as the metastable solubility limit of Ni to liquid Sn is reached. Consequently, the amount of Ni in Sn-based solders is strictly limited to low concentrations. On the other hand, as Ni and Cu are completely soluble in solid-state, Ni can be alloyed to a Cu conductor basically in any proportion. Thus, (ii) the addition of more Ni to Cu conductor (1–2.5 at.%) continues the trend in the thickness ratio of Cu_3Sn to Cu_6Sn_5, produces significant amount of pores at.%t the Cu/Cu_3Sn interface and increases the thickness of Cu_6Sn_5 in comparison to that in the pure Cu/Sn couple. (iii) Further addition of Ni to Cu (around 5 at.%) increases the total thickness of the IMC layer by about two times in comparison to that in the Cu/Sn diffusion couple and makes Cu_3Sn to almost completely disappear (within the resolution limits of SEM). (iv) The addition of 10 of Ni to Cu makes the pores disappear and decreases the total IMC thickness again close to that of the Cu/Sn diffusion couple.

The absence of Cu_3Sn in Cu(Ni) alloys with more than 5 at.% Ni can, in principle, be thermodynamic, kinetic or a combined effect as Ohriner also suggested [17]. At the moment many aspects of the phase equilibria in the ternary Sn–Cu–Ni phase diagram are not unambiguously determined. One reason for the uncertainty concerning the Sn–Cu–Ni phase diagram is that in the low Sn region of the diagram the kinetics are very slow at temperatures relevant to soldering. Thus, even after extensive solid-state annealing, it is not clear whether equilibrium has actually been established. Based on the diagrams calculated by ourselves [22] and those presented in the literature [23–26], there is no obvious reason for the absence of Cu_3Sn in these diffusion couples. To demonstrate this it is appropriate to investigate the metastable isothermal section calculated at 125°C (shown in Fig. 6.10).

Fig. 6.11 Experimentally determined isothermal section of the Sn–Cu–Ni system at 235°C [19]

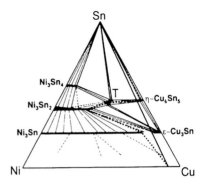

This metastable diagram does not include the solid miscibility gap in the Cu–Ni system nor the ternary phase 45Sn29Cu26Ni (τ) observed in [23]. This is because the formation of τ requires very long annealing times (up to 10,000 h) and, further, the formation of this compound has not been confirmed elsewhere [25]. Two other ternary compounds situated near 25 at.% Sn (τ_1 and τ_2) have been experimentally observed [25]. However, as the formation of these compounds have not been detected at temperatures below 400°C after extensive annealing [25] their formation at 125°C during our experiments is highly unlikely. However, as there is some information that indicates that τ_1 phase is stablized down to room temperature [26] it has been included in the optimisation of the Cu–Ni–Sn diagram (marked as Γ_1). It should also be noted that up to 50% metastable solubility of Ni into the Cu sublattice of Cu_6Sn_5 has been reported at 220°C [27].

Based on the diagram shown in Fig. 6.10, addition of more than 5 at.% of Ni to Cu would lead to local equilibrium between Cu(Ni) and (Ni,Cu)$_3$Sn or ternary compound CuNi$_5$Sn$_2$ (Γ_1). After that, the diffusion path (presented with the dotted line in Fig. 6.10) could go through (Ni,Cu)$_3$Sn$_2$ and then to (Cu,Ni)$_6$Sn$_5$. Thus, in this reaction sequence the Cu$_3$Sn would not be stable and could not form. In addition, as Ni compounds are known to be very slow to grow they could be present as such a thin layers that they cannot be detected within the resolution limits of the FE-SEM. This reaction sequence is, however, quite complicated and it would therefore be difficult to understand why (Cu,Ni)$_6$Sn$_5$ should grow so fast. According to the diagram determined experimentally by Oberndorff at 235°C [23] (Fig. 6.11) the maximum stable solubility of Ni to (Cu,Ni)$_3$Sn is about 3 at.%. The addition of more than 3 at.% of Ni to Cu should lead to diffusion path that would go first to (Ni,Cu)$_3$Sn, then to (Ni,Cu)$_3$Sn$_2$, then through ternary 45Sn29Cu26Ni (τ) and, finally, to (Cu,Ni)$_6$Sn$_5$ and Sn. Thus, Cu$_3$Sn would be absent from the reaction sequence. Again the resulting reaction sequence would be highly complex and does not provide a feasible explanation for the observed phenomena.

Hence, it seems that one has to turn to kinetic considerations to rationalise the absence of Cu$_3$Sn in the Cu(Ni)/Sn diffusion couples with more than 5 at.% of Ni. However, as the chemical potential difference provides the driving force for diffusion, the thermodynamic assessment of the Sn–Cu–Ni system can provide

6.1 Examples of Metal–Metal Interfaces

relevant information about the effect of Ni. As presented by Yu et al. the driving force for diffusion of Sn through Cu_6Sn_5 is increased when the Ni content of this phase is increased [28]. Sn is known to be the main diffusing species in Cu_6Sn_5 whereas Cu is the main diffusing species in Cu_3Sn [18, 29]. Therefore, we will concentrate on the magnitudes of the driving forces for the diffusion of Sn and Cu through Cu_6Sn_5 and Cu_3Sn, respectively. From Fig. 6.10 one can see that when 10 at.% Ni is added to $(Cu,Ni)_6Sn_5$ its stability will increase as already reported earlier [30] and also indicated by first-principle calculations [31]. This will have effect on the chemical potentials (and thereby activities) at the $Sn/(Cu,Ni)_6Sn_5$ and Cu_3Sn/Cu interfaces in such a way that the driving force for Sn diffusion through $(Cu,Ni)_6Sn_5$ increases from ΔG_{Sn} to $\Delta G_{Sn}*$ (almost 4 times ΔG_{Sn}) whereas the driving force for diffusion of Cu through Cu_3Sn decreases from ΔG_{Cu} to $\Delta G_{Cu}*$ (about 3/5 ΔG_{Cu}). Further, if for some reason Cu_3Sn does not form at all, the driving force for Sn diffusion through $(Cu,Ni)_6Sn_5$ will increase even further (dashed tangent line in Fig. 6.12 for $g^{\eta\prime}*$). As the diffusion flux is linearly proportional to the driving force according to the well-known Nernst-Einstein relation, the material flux can be expected to grow over (Cu_6Sn_5) and to decrease over Cu_3Sn. However, it should be emphasised that owing to the lack of diffusion data the discussion above remains qualitative.

At the moment it is not totally clear what microstructural effects Ni induces to the $(Cu,Ni)_6Sn_5$ compound layer. However, what is evident is that the presence of Ni will accelerate Sn diffusion through this layer. An appropriate question is then: is the formation of the pores also related to the changes in the diffusion fluxes inside the Cu_6Sn_5 and Cu_3Sn layers? Van Loo et al. [32] have analysed the effect of intrinsic diffusion coefficients and initial and interfacial concentrations of each phase on the creation and annihilation of vacancies. Any interphase boundary must be able to serve as either source or sink for vacancies to accommodate the differences in vacancy fluxes. In our case, the Cu_6Sn_5/Cu_3Sn interface must be a source of vacancies for both intermetallic compound layers. This is because Sn is the main diffusing species in Cu_6Sn_5 and Cu is the main diffusing species in Cu_3Sn. In order to maintain vacancy flux balance these vacancies, which are created at the Cu_6Sn_5/Cu_3Sn interface must be annihilated at other interfaces or in the bulk phases. Since it is evident that Sn flux through Cu_6Sn_5 is very large it is likely that the vacancies are not accumulated at that side even though their creation will be accelerated at the Cu_6Sn_5/Cu_3Sn interface due to the increased Sn flux. However, because the Cu_6Sn_5/Cu_3Sn interface supplies vacancies to both phases the accelerated diffusion of Sn in the Cu_6Sn_5 layer could cause extra vacancies to be injected also into the Cu_3Sn layer. When the driving force for the Cu diffusion through Cu_3Sn decreases it is also likely that its flux decreases. In fact, the diffusion flux is bound to decrease if the interdiffusion coefficient in $(Cu,Ni)_3Sn$ is decreased at the same time. This could lead to the accumulation of vacancies at the Cu_3Sn/Cu interface and eventually to the formation of macroscopic voids. Especially, if we assume that the ratio of mobilities of Cu and Sn in $(Cu,Ni)_3Sn$ does not markedly change due to the addition of Ni, the Kirkendall

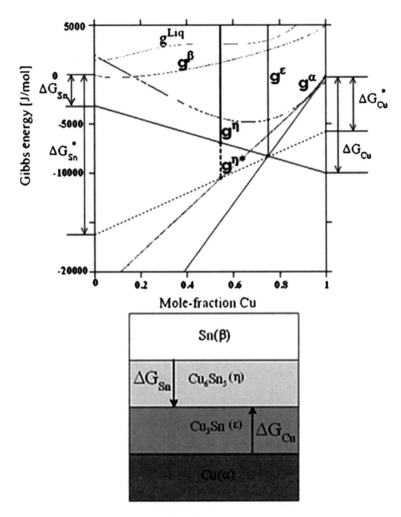

Fig. 6.12 Gibbs free energy diagram at 125°C showing the driving forces for diffusion over the schematically represented interfacial reaction zone

effect should be enhanced, as the interdiffusion coefficient and therefore material flux through (Cu,Ni)$_3$Sn is decreased. As more Ni is added to Cu the interdiffusion coefficient in (Cu,Ni)$_3$Sn would be further and further decreased (along with the driving force for Cu diffusion), finally resulting into (almost) total disappearance of (Cu,Ni)$_3$Sn as well as the pores. Thus, the changes in the diffusion fluxes owing to the Ni addition (at least partly due to the changes in the driving forces) could result both in the observed thickness alterations of IMC layers as well as in the formation and disappearance of pores.

6.1 Examples of Metal–Metal Interfaces

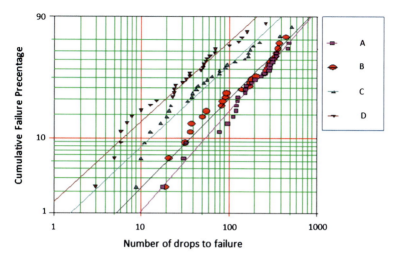

Fig. 6.13 Weibull reliability plots from drop test results of assemblies with the Ni(P)|Au-finished soldering pads having four different P–chemistries

6.1.3 Ni or Ni(P) Metallisation on PWB Versus SnAgCu solder

During recent decades Ni/Au coatings have been extensively used in high-density component assemblies as a surface finish on printed wiring boards (PWB) because it has many advantages over other surface finishes such as hot air solder levelling (HASL) Sn finishes, as discussed above. Further, Ni/Au coatings provide higher mechanical strength, hardness and resistance against thermal fatigue of lead-free solder interconnections than can be achieved when using organic solder preservatives (OSP), Sn or Ag on Cu pads [33].

In addition to the above-mentioned benefits, there are also a few typical reliability problems related to interfacial reactions between Ni/Au coatings with Sn-based solders. Especially, the usage of electroless Ni/immersion Au (ENIG) finishes has lead to reduced reliability [34–42], because during the electroless plating process phosphorus is co-precipitated with Ni. It is the presence of phosphorus in the surface finish layers that has been observed to be associated with severe reliability problems. Although the wetting occurs properly and the chemical reaction between Sn and Ni is evident, the interfacial strength is not adequate. The weakest interfacial reaction product readily fractures under mechanical stress and leaves behind an open circuit

Figure 6.13 shows the results from the drop test carried out according to the JESD22-B111 standard. The component was SnAgCu bumped (144 bumps having diameter 500 μm) CSP-component (12 mm*12 mm), the solder was near-eutectic SnAgCu and four different ENIG coatings on Cu pad was used. The phosphorus content of the coatings A and B was ∼8 wt%, coating C had over 10 wt% phosphorus and D had less than 3 wt% P The characteristic lifetimes (η) of the

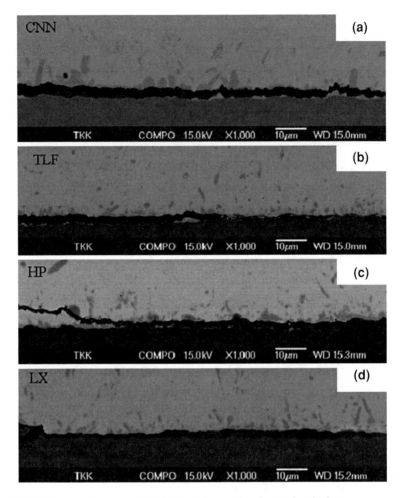

Fig. 6.14 Failure mechanisms related to the different phosphorus chemistries

coatings were $\eta_A = 286$, $\eta_B = 240$, $\eta_C = 101$ and $\eta_D = 50$ drops to failure. As can be seen from Fig. 6.14 the failure mechanisms related to the different phosphorus chemistries do not indicate detectable differences between the A, B and C. However, the low phosphorus coating (D) seems to be the weakest as no reaction products or solder can be seen between the crack and the coating.

The root cause for the brittle fracture has been discussed in many papers dealing with the reactions between electroless Ni and SnPb-solders as well as lead-free solders, mainly near-eutectic SnAgCu [43–53]. However, the formation mechanism of the interfacial reaction products that causes the reliability problems is not yet thoroughly understood. It has been proposed that the segregation of phosphorus on the PWB side of the fractured surface is responsible for the failure at the $Ni_3Sn_4/Ni(P)$ interface in as-reflowed BGA (Ball Grid Array) solder

6.1 Examples of Metal–Metal Interfaces

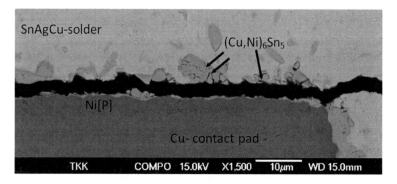

Fig. 6.15 Brittle fracture (after drop test) in SnAgCu- solder/Ni[P] interface

Fig. 6.16 SEM micrograph revealing the presence of two reaction layers between the IMC and Ni[P] coating

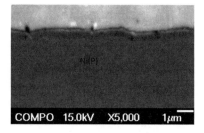

interconnections [41]. Jang et al. [54] detected phosphorus enrichment in the Si/SiO$_2$/Al/Ni(P)/63Sn37Pb multilayer structure after reflow and suggested that the formation mechanism of the interfacial reaction layers is so-called solder assisted crystallisation. The mechanism is based on the preferential dissolution of Ni from the Ni(P) layer, which leads to an increase in the phosphorus content of the upper part of the Ni(P) layer and the subsequent formation of Ni$_3$P [54]. There are also suggestions that the brittle fracture is related to the redeposition of AuSn$_4$ at the interface after high temperature annealing [6, 14, 54–58]. However, this can only happen if the amount of Au (i.e. the thickness of the Au coating) exceeds its solubility in β-Sn at the annealing temperature [59]. In particular, when the intermetallic compound (IMC) formed at the interface after reflow is Cu$_6$Sn$_5$ or (Cu,Ni)$_6$Sn$_5$, (in the case of Cu-bearing solders), the latter explanation is not plausible.

Figure 6.15 shows the interfacial microstructure between the NiP and near eutectic SnAgCu solder from the sample that has been under mechanical shock loading (drop test) after the assembly reflow. The point analysis taken from the interfacial IMC layer gives a composition of 40 at.% Cu, 15 at.% Ni and 45-at.% Sn indicating that the IMC is (Cu,Ni)$_6$Sn$_5$, where Ni is dissolved into the Cu sublattice. Similar results have been reported in several papers dealing with the Sn–Cu–Ni system [60–63].

Between the IMC and Ni[P] coating two layers having different contrast can be observed (Fig. 6.16). Next to Ni[P] there is a thin (less than 0.5 μm) dark layer and

Fig. 6.17 Bright-field TEM image of the interfacial region in the sample that have been reflowed five times

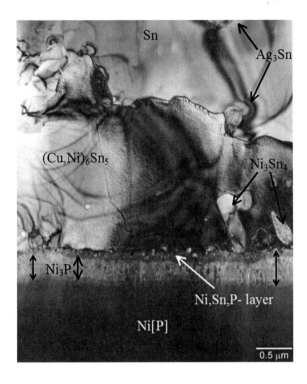

on the top of that a very thin layer with lighter contrast. Both layers form rapidly, since they are visible already after the soldering. The contrast of the topmost layer indicates that it contains heavier elements than in the other two layers below. Figure 6.17 taken from the sample that has been reflowed five times shows a bright-field TEM image from the IMC and phosphorus enriched area. From the TEM micrograph it is confirmed that the layer between the electroless Ni[P] and the $(Cu,Ni)_6Sn_5$ is not a single reaction layer but is composed of two layers, as indicated also by the SEM micrograph (Fig. 6.16). The layer next to the electroless Ni[P] is a crystalline Ni_3P. The first time this phase was observed by Jang et al. was when studying the $Si/SiO_2/Al/Ni(P)/63Sn37Pb$ multilayer structure after the reflow [54].

As shown in Fig. 6.17, the crystalline Ni_3P on top of the original Ni[P] coating has a columnar structure in which some organic impurities and tin seem to be concentrated between the columns (bright and dark stripes). It is to be noted that there are a few Ni_3Sn_4 crystals inside the $(Cu,Ni)_6Sn_5$ close to the ternary layer and some small Ag_3Sn precipitates inside the Sn matrix (Fig 6.17). Figure 6.18 shows high resolution EM (HREM) micrograph from the layer between $(Cu,Ni)_6Sn_5$ and Ni_3P. The composition of this phase is estimated to be 55 at.% Ni, 35 at.% Sn and 10 at.% P and it has a nanocrystalline structure, based on the selected area diffraction (SAD) results (not shown here). The nanocrystalline (or partly amorphous) nature of the ternary layer is also evident from the HREM micrograph as no

6.1 Examples of Metal–Metal Interfaces 155

Fig. 6.18 HREM image of the organic particles inside the nanocrystalline Ni–Sn–P layer

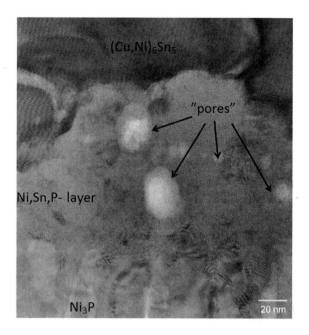

traces of lattice planes can be seen inside the ternary layer. There are some crystalline areas at the vicinity of the Ni$_3$P layer that most probably are small Sn grains. It is unlikely that this phase is the Ni$_2$SnP[64] or Ni$_3$SnP[52, 56] reported earlier. The ternary NiSnP layer contains numerous small defects that have a pore-like appearance as can be seen from Figs. 6.17 and 6.18.

It is likely that these structural defects assists the crack propagation in the layer and thus makes interconnections behave in an extremely brittle manner. The "pores" were also observed by Jeon et al. [65] in Ni[P] | (SnPb)$_{eut}$ reactions and they claim that they are Kirkendall voids. It is difficult to accept that Kirkendall voids would form it the interfacial region in such a short time as they are caused by the unbalanced intrinsic fluxes of materials in solid-state. Thus, the formation of Kirkendall voids typically requires time. Further, we did not observe any detectable change in the number or size of these "pores" in the ternary NiSnP layer when comparing samples that were reflow-soldered five times and annealed at 170°C for 64 h after the reflow (see Fig. 6.19). In fact, Fig. 6.18 points out that the pores have an internal structure and contain organic material that most probably originates from the Ni(P) plating bath.

When attempting to rationalise the formation of the observed reaction structures with the help of the solder-assisted crystallisation approach, we face several problems. The presence of a nanocrystalline NiSnP layer between the Ni$_3$P and IMC cannot be explained by preferential dissolution of Ni from Ni[P] to liquid solder. This would require notable solid-state diffusion of Ni inside the Ni[P] coating so that Ni atoms can reach the Ni[P]/solder interface. This can hardly take place during the short soldering process, i.e. within less than a minute. Further, the

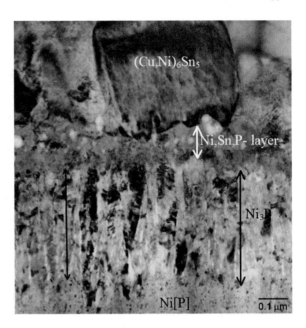

Fig. 6.19 Bright field image from the reaction zone in the sample that has been annealed at 170°C for 64 h after the assembly reflow

solubility of P to liquid Sn (and Sn-based solders) is much larger than that of Ni [66]. As the solubility is directly related to the dissolution rate of the given material, the dissolution rate of P to liquid Sn should consequently be higher than that of Ni. Therefore, the solder-assisted crystallisation approach cannot be utilised to the more complex lead-free system considered here.

In order to explain the formation of the above-mentioned reaction products more quantitatively the thermodynamic description of the quaternary Sn–Ni–P–Cu system is needed, for Ag can be ignored in the analysis, as it has only minor effect on the reaction through activity. Unfortunately, there are not enough reliable data on the quaternary system as would be required for its critical thermodynamic assessment. However, it is still possible to rationalise the reaction sequence by making use of the available binary phase diagrams, Sn–P and Ni–P [67], as well as thermodynamic evaluations concerning the stability of ternary Sn–P–Ni liquids [68]. In addition, there exists a very recent experimental investigation on the Sn–P–Ni system, which also includes the determination of a new ternary phase [69]. However, these results are determined at high temperature (700°C and above) with equilibrated bulk materials, which are very different from the actual chemically deposited Ni(P) metallisation, and represent global equilibrium conditions. As in soldering, we are typically dealing with local (metastable) equilibrium and the temperatures are much lower the experimental results presented in [69] do not offer any help in the analysis.

The reaction between Ni(P) and liquid SnAgCu solder alloy starts with instant dissolution of a thin (flash) Au layer, which is followed by the dissolution of Ni and P. Due to the supersaturation of phosphorus in the Sn-rich solder a thin layer of new liquid (L2) is formed between the solid Ni(P) and the bulk liquid solder (L1). This argument is supported by the presence of a liquid miscibility gap in the binary Sn-P system that must extend to the ternary Sn–P–Ni system and can also continue to lower temperatures as a metastable miscibility gap [68]. In fact our preliminary calculations indicate that the metastable miscibility gap is strongly stabilized by dissolved Ni. When the two liquids come in local equilibrium with each other as well as with the Ni(P) substrate, Ni, P and Sn will redistribute between the liquids so that the liquid L2 contains large amounts of Ni and P and a small amount of Sn, while the Sn-rich liquid L1 has some Ni and a small amount of P. Because of the high P and Ni contents, the liquid L2 must be unstable at these low temperatures and, therefore, it turns rapidly to a nanocrystalline ternary layer providing a solid substrate for the $(Cu,Ni)_6Sn_5$ to nucleate and grow.

During cooling after the assembly reflow or during successive reflows the nanocrystalline NiSnP layer partly transforms into columnar Ni_3P and the extra atoms, not being used in the formation of Ni_3P, are rejected to the ternary NiSnP layer. In addition to P and Sn, organic additives that are always present in electroless Ni(P) coatings also precipitate out at the interfaces between the columnar Ni_3P crystals, as well as in the ternary NiSnP phase, where they are revealed as numerous small "pores", i.e. organic impurity particles, as shown in Figs. 6.17 and 6.18. Therefore, the Kirkendall voids reported earlier may, in fact, be organic particles. Also the white stripes between the columnar Ni_3P crystals in Fig. 6.17 as observed also by Matsuki, Ikuba, and Saka [64] are most likely organic material.

It is interesting to find out that when the P content of the electroless Ni(P) coating is high enough, the formation of Ni_3P is suppressed. This is shown in Fig. 6.20a, where Ni(P) having a high (30 at.%) phosphorus content has reacted with the near-eutectic Sn–Ag–Cu solder during reflow. Only the ternary NiSnP (τ) reaction layer is visible between Ni(P) and the solder. In addition, the brittle fracture caused by mechanical shock loading has propagated at the interface between Ni[P] and the ternary layer. Furthermore, extensive spalling of two intermetallics $(Cu,Ni)_6Sn_5$ and $(Ni,Cu)_3Sn_4$, being in local equilibrium, can be seen. When the soldering time was extended to 20 min the difference between high (30 at.%) P content Ni[P] component under bump metallisation and lower (14 at.%) P content Ni[P] PWB coating is even more evident (see Fig 6.20b and c). As the time above liquidus is now longer, more Ni (and P) is dissolved in the solder and the $(Cu,Ni)_6Sn_5$ near the component side interface is totally transformed to $(Ni,Cu)_3Sn_4$ as a result of the limited supply of Cu. It is also interesting to note from Fig. 6.20 that $(Ni,Cu)_3Sn_4$ is never found in contact with the τ-phase. If $(Cu,Ni)_6Sn_5$ is present in the system, it is always located between the τ-phase and $(Ni,Cu)_3Sn_4$. When $(Cu,Ni)_6Sn_5$ is transformed into $(Ni,Cu)_3Sn_4$ (Fig. 6.20b), the $(Ni,Cu)_3Sn_4$ phase is only found inside the bulk solder matrix away from the interface. This can be taken as an indication that $(Ni,Cu)_3Sn_4$ cannot exist in local equilibrium with the τ-phase whereas $(Cu,Ni)_6Sn_5$ can.

Based on the above experimental result it seems that the miscibility gap extends to the ternary systems in such a way that a higher P content in the

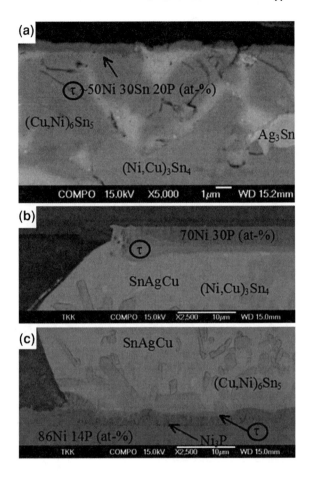

Fig. 6.20 Backscattered SEM images **a** the component side, where the P concentration in Ni(P) is about 30 at.% annealed for 5 min at 250°C, **b** the component side annealed for 20 min at 250°C and, **c** the PWB side, where the P concentration in Ni(P) is about 14 at.%, annealed for 20 min at 250°C

original Ni(P) also results in a higher P content in the liquid L2 making it not possible for Ni_3P to form. This indicates that the higher phosphorus content will further stabilize the nanocrystalline NiSnP phase at the expense of the crystalline Ni_3P, mainly for the following reason. It is known that the crystallisation (and also the glass transition) temperature of an amorphous phase is usually lower at the eutectic point (19 at.% P in the Ni–P system) than near intermediate compounds or end-elements [70]. Therefore, we can expect that an increase in the P content of the nanocrystalline NiSnP layer will also increase its crystallisation temperature, as the composition is shifted further away from the eutectic point. Furthermore, support for the effect of P on the stability of amorphous structures can be found from the literature, where it has been widely documented that when processing diffusion barriers for thin film applications amorphous structures are frequently realised by alloying elements such as B, C, N, Si and P with transition metals [71] (see also Sect. 4.3.3).

6.1.4 Redeposition of AuSn$_4$ on Top of Ni$_3$Sn$_4$

Gold is generally used in electronics as a thin protective layer to ensure the solderability of the underlying layers. Thus, the amount of Au present in the solder interconnections is usually quite small. However, the behaviour of these small amounts of Au with other metals is theoretically interesting and is of great importance in soldering applications. One of the most important interactions is the interplay between Ni and Au during prolonged solid-state annealing. Over 10 years ago Mei et. al [41]. revealed a problem that was peculiar to the Ni/Au metallisation with SnPb solders. They discovered that after prolonged ageing (150°C for 2 weeks) the AuSn$_4$ intermetallic compound, which had formed during the soldering in the bulk solder, redeposited at the solder/substrate interface. The reconstituted interface was significantly weakened and failed by brittle fracture along the surface between the redeposited AuSn$_4$ and the Ni$_3$Sn$_4$ layer formed during the reflow.

Minor et.al. [72] subsequently studied the mechanism of the redeposition of the AuSn$_4$ intermetallic compound. They discovered that the as-solidified solder interconnections contained dense distributions of small needle-like AuSn$_4$ particles evenly distributed throughout the bulk solder. The interface between Ni and the bulk solder consisted of the layer of Ni$_3$Sn$_4$ that contained a very small amount of Au. A coarse intermetallic layer developed above the Ni metallisation during the aging. It thickened roughly as $t^{0.5}$, i.e. indicating that the growth was diffusion controlled. The simultaneous depletion of AuSn$_4$ needles from the bulk occurred. The redeposited intermetallic compound in the aged samples appeared to have a composition close to AuSn$_4$.

Ho et.al. [73] investigated how the addition of small amounts of Cu into SnPb solder would influence the redeposition behaviour of AuSn$_4$. After the reflow, the gold was completely absent from the interface and all the AuSn$_4$ intermetallic particles were evenly distributed throughout the bulk solder. At the Ni/solder interface a Ni$_3$Sn$_4$ layer was formed except for the sample where solder included Cu. In this case the interfacial reaction layer was not Ni$_3$Sn$_4$ but Au-bearing (Cu$_{1-p-q}$Au$_p$Ni$_q$)$_6$Sn$_5$ quaternary compound. After the solid-state annealing redeposition of AuSn$_4$ on top of Ni$_3$Sn$_4$ took place; however when the interfacial IMC was (Cu$_{1-p-q}$Au$_p$Ni$_q$)$_6$Sn$_5$, no traces of AuSn$_4$ were detected.

Shiau et.al [74]. carried out a study on reactions between lead-free Sn–Ag–Cu solder and an Au/Ni finish in order to find out whether the redeposition of AuSn$_4$ intermetallic compound would take place also in this lead-free system. Three solders were used: Sn3.5Ag, Sn4Ag0.5Cu and Sn3.5Ag0.75Cu. In the Sn3.5Ag alloy a thin layer of Ni$_3$Sn$_4$ had formed at the interface, while in the Sn3.5Ag0.75Cu a layer of (Cu$_{1-p-q}$Au$_p$Ni$_q$)$_6$Sn$_5$ was formed after reflow. In the Sn4Ag0.5Cu solder both Ni$_3$Sn$_4$ and (Cu$_{1-p-q}$Au$_p$Ni$_q$)$_6$Sn$_5$ were present near the interface. The formation of different IMC layers as a function of Cu concentration in the solder is readily rationalised with the concept of critical Cu concentration discussed in detail above. In the case of the Cu-free solder, minor redeposition of (Au,Ni)Sn$_4$ was observed at the interface. It should be noted that, once again, in the systems with Cu no AuSn$_4$ redeposition occurred.

Fig. 6.21 SEM micrograph from the (SnPbAg)/Ni/Au system after annealing for 250 h at 150°C

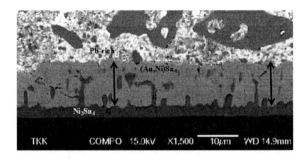

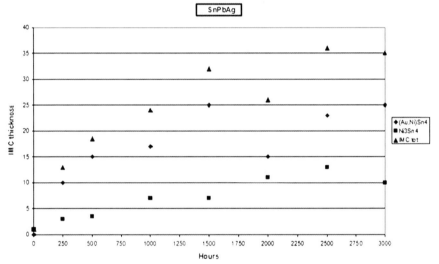

Fig. 6.22 Plot of the IMC thickness versus time in the (SnPbAg)/Ni/Au system

Next we will present our own experimental results obtained with three different solders [Sn36.0Pb2.0Ag, Sn3.5Ag and Sn3.8Ag0.7Cu (wt%)] and a 5 μm-thick Au layer on top of Ni metallisation. Details of the experimental set-up and results can be found from [59]. After annealing for 250 h at 150°C the $(Au,Ni)Sn_4$ redeposition can be clearly seen to have taken place in the (SnPbAg)/Ni/Au system (Fig. 6.21). The thickness of the $(Au,Ni)Sn_4$ is remarkably large and also numerous cracks visible in the layer indicate that it is most probably brittle. Likewise, the interface between Ni_3Sn_4 and $(Au,Ni)Sn_4$ shows some cracking (Fig. 6.21). It should be emphasised that the excessive thickness of the $(Au,Ni)Sn_4$ is related to the thick Au coating used in this study. Thus, the results of this study should not be compared directly to the earlier investigations where much thinner Au-layers were used [72–75]. Nevertheless, it is equally important to realise that the reaction mechanisms themselves are most likely identical. The formation of a Pb-rich layer in front of the redeposited $(Au,Ni)Sn_4$ intermetallic compound is also evident from the figure. The growth of the IMC layers with respect to time is shown in Fig. 6.22.

6.1 Examples of Metal–Metal Interfaces

Table 6.1 Minimum (d_{min}), maximum (d_{max}) and average thickness (d_{av}) values, compositions as well as the growth exponents (n) determined by the regression analysis for the phases formed in (SnPbAg)/Ni/Au reaction couple

Time@150°C (h)	$(Ni_x,Au_{1-x})_3Sn_4$ thickness (µm)	n	$(Au_y,Ni_{1-y})Sn_4$ thickness (µm)	n	IMC total thickness (µm)	n
After reflow	$d < 1$		Not observed		$d \sim 1$	
250	$d_{av} = 3$ $d_{min} = 2.5$ $d_{max} = 3.5$ $x = 1$	0.7	$d_{av} = 10$ $d_{min} = 6$ $d_{max} = 11$ $y = 0.5$	0.5	$d_{av} = 13$ $d_{min} = 9$ $d_{max} = 14$	0.5
500	$d_{av} = 3,5$ $d_{min} = 2.5$ $d_{max} = 4$ $x = 1$	0.7	$d_{av} = 15$ $d_{min} = 13$ $d_{max} = 18$ $y = 0.5$	0.5	$d_{av} = 19$ $d_{min} = 16$ $d_{max} = 22$	0.5
1,000	$d_{av} = 7$ $d_{min} = 6$ $d_{max} = 8$ $x = 1$	0.7	$d_{av} = 17$ $d_{min} = 14$ $d_{max} = 25$ $y = 0.5$	0.5	$d_{av} = 24$ $d_{min} = 20$ $d_{max} = 35$	0.5
1,500	$d_{av} = 7$ $d_{min} = 6$ $d_{max} = 8$ $x = 1$	~ 0	$d_{av} = 25$ $d_{min} = 23$ $d_{max} = 32$ $y = 0.5$	0.5	$d_{av} = 32$ $d_{min} = 30$ $d_{max} = 40$	0.5
2,000	$d_{av} = 11$ $d_{min} = 10$ $d_{max} = 12$		$d_{av} = 15$ $d_{min} = 13$ $d_{max} = 17$		$d_{av} = 26$ $d_{min} = 24$ $d_{max} = 27$	
2,500	$d_{av} = 13$ $d_{min} = 10$ $d_{max} = 14$		$d_{av} = 23$ $d_{min} = 20$ $d_{max} = 25$		$d_{av} = 36$ $d_{min} = 30$ $d_{max} = 40$	
3,000	$d_{av} = 10$ $d_{min} = 9$ $d_{max} = 12$		$d_{av} = 25$ $d_{min} = 19$ $d_{max} = 29$		$d_{av} = 35$ $d_{min} = 30$ $d_{max} = 40$	

The corresponding minimum and maximum thickness values, compositions as well as the growth exponents determined by the regression analysis for the phases formed as a function of time are shown in Table 6.1.

The total IMC layer grows more or less with parabolic kinetics (n ≈ 0.5). As can be seen from Table 6.1 the growth of $(Au,Ni)Sn_4$ is also diffusion controlled (the growth exponent n ≈ 0.5). The growth kinetics of Ni_3Sn_4 appears to be somewhat different as n ≈ 0.7. The thickness of the Ni_3Sn_4 layer also remains small and the IMC seems to have achieved its limiting value after 1,000 h of annealing as the average thickness remains the same (after 1,500 h) and the growth exponent seems to approach zero (Fig. 6.22 and Table 6.1). However, when the samples are further annealed the Ni_3Sn_4 continues to grow with more or less parabolic type kinetics and after 3,000 h the thickness of Ni_3Sn_4 is about 10 µm (that of $(Au,Ni)Sn_4$ is about 25 µm). The growth exponents for longer annealing times are not reported in Table 6.1 as there is a notable drop in the thickness of the IMCs after 2,000 h after which the growth seems to resume again.

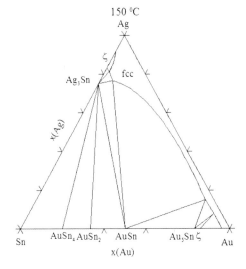

Fig. 6.23 Isothermal section at 150°C from the evaluated Ag–Au–Sn ternary phase diagram

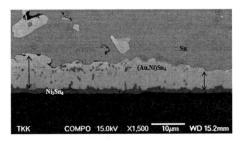

Fig. 6.24 SEM micrograph from the (SnAg)/Ni/Au system after annealing for 250 h at 150°C

This is most probably caused by the local variation in the thickness of the IMC's together with limited experimental accuracy. The Ni content of the (Au,Ni)Sn$_4$ was about 10 at.% [i.e. 50% of Au atoms in the Au-sublattice have been replaced by Ni atoms (Table 6.1)]. In contrast, no Au could be detected inside Ni$_3$Sn$_4$ within the resolution limits of EDS. Likewise, no Ag was found in any of the above-mentioned IMC layers. It was detected only in the form of an Ag$_3$Sn compound throughout the solder matrix. Despite the fact that Ag$_3$Sn can exist in local equilibrium with AuSn$_4$ (Fig. 6.23) at 150°C it was not found at the interface. However, inside the bulk solder matrix Ag$_3$Sn and AuSn$_4$ precipitates were commonly found to be in contact.

The (SnAg)/Ni/Au system shows similar behaviour as the (SnPbAg)/Ni/Au system. The interfacial structure after annealing at 150°C for 250 h is shown in Fig. 6.24 The corresponding IMC thickness versus time is shown in Fig. 6.25. Again the corresponding minimum/maximum thickness values, compositions as well as the growth exponents determined by the regression analysis for the phases formed as a function of time are shown in Table 6.2.

The total IMC layer grows with parabolic kinetics up to 500 h (n ≈ 0.5). However, around 1,000 h the growth exponent of the total IMC layer becomes

6.1 Examples of Metal–Metal Interfaces

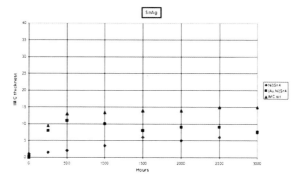

Fig. 6.25 Plot of the IMC thickness versus time in the (SnAg)/Ni/Au system

zero and thus indicates that the limiting thickness of the layer has been reached. No further growth of total IMC layer thickness is observed up to 3,000 h of annealing. The growth kinetics of (Au,Ni)Sn$_4$ are also parabolic (n ≈ 0.5) (Table 6.2). The thickness of the (Au,Ni)Sn$_4$ is less than half of that in the (SnPbAg)/Ni/Au system and it seems that the limiting thickness is achieved around 500 h. The time is the same when also the total IMC thickness ceases to increase. (Table 6.2). In fact, after 1,000 h of annealing the average thickness of the (Au,Ni)Sn$_4$ phase starts to decrease and the growth exponent becomes negative. This trend slowly continues when annealing times are extended to 3,000 h. The growth of Ni$_3$Sn$_4$ seems to follow the linear kinetics whereas in the SnPbAg case the growth seemed to be a mixture of diffusion and reaction control.

It should be noted that the average thickness of the (Ni,Au)$_3$Sn$_4$ continues to increase after 1,000 h whereas that of (Au,Ni)Sn$_4$ decreases and the growth exponents exceeds one. This indicates that the Ni$_3$Sn$_4$ layer is partly growing at the expense of the (Au,Ni)Sn$_4$ layer. The Ni content of the (Au,Ni)Sn$_4$ is again around 10 at.%. On the contrary, no Au can be detected inside Ni$_3$Sn$_4$ (Table 6.2). After 2,000 h the Ni layer seems to be totally consumed and (Cu,Ni,Au)$_6$Sn$_5$ is formed between (Au,Ni)Sn$_4$ and (Ni,Cu)$_3$Sn$_4$ indicating that the underlying Cu starts to participate in the reactions. This (Cu,Ni,Au)$_6$Sn$_5$ layer continues to grow slowly between the (Au,Ni)Sn$_4$ and (Ni,Cu)$_3$Sn$_4$ phases as shown in Fig. 6.26. Because Cu starts to take part in the reactions, the growth exponents for the longer annealing times could not be determined.

In comparison to the (SnPbAg)/Ni/Au and (SnAg)/Ni/Au systems, the (SnAgCu)/Ni/Au system behaves quite differently. From Fig. 6.27 it can be seen that no redeposition of AuSn$_4$ intermetallic compound takes place after annealing at 150°C for 250 h (or up to 1,500 h). The morphology of the structure shown in the micrograph (Fig. 6.27) and the compositional analyses indicated that there are two reaction product layers with different Au to Ni ratios. The top layer is (Cu,Au,Ni)$_6$Sn$_5$, where the Au content is about 12 at.%, and the Ni content is about 2–3 atomic percents. The lower (Cu,Ni,Au)$_6$Sn$_5$ layer contains only about 5 at.% Au and about 17 at.% Ni. Some of the Ni signal may arrive from the underlying Ni substrate owing to the small thickness of the lower reaction layer.

Table 6.2 Minimum (d_{min}), maximum(d_{max}) and average thickness (d_{av}) values, compositions as well as the growth exponents (n) determined by the regression analysis for the phases formed in (SnAg)/Ni/Au reaction couple

Time@150°C (h)	$(Ni_x,Au_{1-x})_3Sn_4$ thickness d (µm)	n	$(Au_y,Ni_{1-y})Sn_4$ thickness (µm)	n	IMC total thickness (µm)	n
After reflow	$d \sim 1$		Not observed		$d \sim 1.5$	
250	$d_{av} = 1,5$ $d_{min} = 1$ $d_{max} = 2$ $x = 1$	1	$d_{av} = 8$ $d_{min} = 5$ $d_{max} = 11$ $y = 0.5$	0,5	$d_{av} = 9$ $d_{min} = 7$ $d_{max} = 12$	0.5
500	$d_{av} = 2$ $d_{min} = 1.5$ $d_{max} = 2.5$ $x = 1$	1	$d_{av} = 11$ $d_{min} = 10$ $d_{max} = 12$ $y = 0.5$	0,5	$d_{av} = 13$ $d_{min} = 11$ $d_{max} = 14$	0.5
1,000	$d_{av} = 3.5$ $d_{min} = 2$ $d_{max} = 5$ $x = 1$ (some $\sim$6–8at.% Cu detected)	>1	$d_{av} = 10$ $d_{min} = 6$ $d_{max} = 13$ $y = 0.5$	<0	$d_{av} = 13.5$ $d_{min} = 10$ $d_{max} = 16$	~ 0
1,500	$d_{av} = 6$ $d_{min} = 5$ $d_{max} = 7$ $x = 1$	>1	$d_{av} = 8$ $d_{min} = 6$ $d_{max} = 9$ $y = 0.5$	< 0	$d_{av} = 14$ $d_{min} = 11$ $d_{max} = 15$	~ 0
2,000	$d_{av} = 5$ $d_{min} = 4$ $d_{max} = 7$ $x = 1$ (Ni layer consumed and $(Cu,Ni,Au)_6Sn_5$??? Formed between the phases		$d_{av} = 9$ $d_{min} = 8$ $d_{max} = 10$ $y = 0.5$		$d_{av} = 14$ $d_{min} = 12$ $d_{max} = 15$	
2,500	$d_{av} = 6$ $d_{min} = 4.5$ $d_{max} = 7.5$ $x = 1$ (Ni layer consumed and $(Cu,Ni,Au)_6Sn_5$??? Formed between the phases		$d_{av} = 9$ $d_{min} = 5$ $d_{max} = 15$ $y = 0.5$		$d_{av} = 14$ $d_{min} = 11$ $d_{max} = 20$	
3,000	$d_{av} = 7,5$ $d_{min} = 6$ $d_{max} = 8$ $x = 1$ (Ni layer consumed and $(Cu,Ni,Au)_6Sn_5$??? Formed between the phases		$d_{av} = 7,5$ $d_{min} = 10$ $d_{max} = 13$ $y = 0.5$		$d_{av} = 15$ $d_{min} = 13$ $d_{max} = 20$	

6.1 Examples of Metal–Metal Interfaces

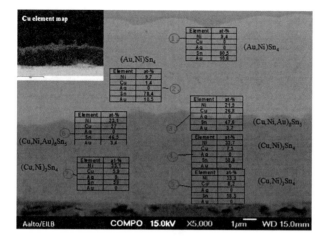

Fig. 6.26 SEM micrograph together with point analyses and Cu element map from the (SnAg)/Ni/Au system after annealing for 2000 hours at 150°C

Fig. 6.27 SEM micrograph from the (SnAgCu)/Ni/Au system after annealing for 250 h at 150°C

The total IMC thickness (Fig. 6.28) is also small when compared to the (SnAg)/Ni/Au and (SnPbAg)/Ni/Au systems (Figs. 6.22 and 6.25) achieving about 7–8 μm after 3,000 h of annealing. It is to be noticed that the y-axis in Fig. 6.28 has a different scale to those in Figs. 6.22 and 6.25. The corresponding minimum/maximum thickness values and compositions for the phases formed as a function of time are shown in Table 6.3. As the thickness increase of the IMCs in this system after 250 h of annealing was almost negligible (Table 6.3), we could not determine the growth exponents for this case. Based on the results obtained, it seems that the growth of the Cu_6Sn_5 is restricted by the limited supply of Cu in the solder. This is also the most probable explanation for the experimental observations made by Ho et.al. [73].

The first and foremost difference between the three systems investigated is that in both (SnAgPb)/Ni/Au and (SnAg)/Ni/Au systems the first phase to form is Ni_3Sn_4, whereas in the (SnAgCu)/Ni/Au system the first phase is $(Cu,Ni,Au)_6Sn_5$. Since this is strongly related to the differences observed in the redeposition behaviour of $AuSn_4$ as $(Au,Ni)Sn_4$, it is important to know why the formation of

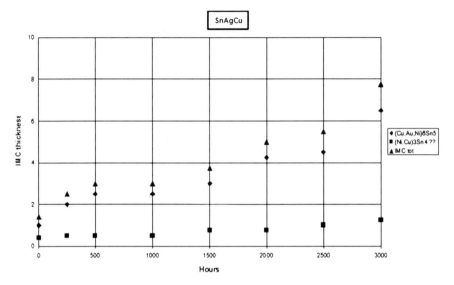

Fig. 6.28 Plot of the IMC thickness versus time in the (SnAgCu)/Ni/Au system

(Cu,Ni,Au)$_6$Sn$_5$ takes place instead of Ni$_3$Sn$_4$ in the (SnAgCu)/Ni/Au system. This issue was addressed in detail above and is thus not repeated here.

In order to analyse the reasons for the redeposition of the AuSn$_4$ all the appropriate interfaces formed in the reaction couples, i.e. Ni$_3$Sn$_4$/AuSn$_4$, Cu$_6$Sn$_5$/AuSn$_4$ and AuSn$_4$/solder, must be investigated to check that they fulfill the requirement of local equilibrium with each other. The (Au,Ni)Sn$_4$/solder interface will be investigated first.

The solder side is in all cases either SnPb (Fig. 6.21) or practically pure tin (Fig. 6.25). In the case of almost pure tin the local equilibrium is fulfilled, as (Au,Ni)Sn$_4$ can exist in local equilibrium with tin. The more interesting case is the SnPb (or in our case SnPbAg) solder as it can be seen that the Pb-rich layer is preferentially located next to (Au,Ni)Sn$_4$ layer (Fig. 6.21). From Fig. 6.29 one can see that lead, with about 20 at.% tin dissolved in it, can exist in local equilibrium with AuSn$_4$.

Therefore also (Au,Ni)Sn$_4$ can most probably exist in local equilibrium with Pb saturated with Sn, thus explaining why the Pb-rich layer can be located beside the (Au,Ni)Sn$_4$ layer. The preferential location of the Pb-rich layer in contact with (Au,Ni)Sn$_4$ has sometimes been explained with the help of interfacial energies [76] as it is known that Pb can form low energy interfaces [77]. Nevertheless, as no reliable experimental values for the surface energies of solids exist, the explanation is highly qualitative. The more probable reason is that tin has been consumed during the reactions with Au and Ni and the Pb-rich layer has been left behind. It should be noticed that the layer cannot be pure lead but instead needs to be saturated with Sn as only Pb(Sn) can exist in local equilibrium with AuSn$_4$ (and also probably with (Au,Ni)Sn$_4$). This also explains why Au–Pb IMCs are not

6.1 Examples of Metal–Metal Interfaces

Table 6.3 Minimum (d_{min}), maximum (d_{max}) and average thickness (d_{av}) values and compositions for the phases formed in (SnAgCu)/Ni/Au reaction couple

Time@150°C (h)	$(Cu_{1-p-q}Au_pNi_q)_6Sn_5$ thickness (μm)	$(Cu_{1-r-s}Ni_rAu_s)_6Sn_5$ thickness (μm)	IMC 2 thickness (μm)	IMC total thickness (μm)
After reflow	$d \sim 1$			$d \sim 1$
250	$d_{av} = 2$ $d_{min} = 1.5$ $d_{max} = 4$ $p = 0.25$ and $q < 0.1$	$d < 1$ $r = 0.3$ and $s < 0.1$		$d_{av} = 2.5$ $d_{min} = 2$ $d_{max} = 5$
500	$d_{av} = 2.5$ $d_{min} = 2$ $d_{max} = 4$ $p = 0.25$ and $q < 0.1$	$d < 1$ $r = 0.3$ and $s < 0.1$		$d_{av} = 3$ $d_{min} = 2.5$ $d_{max} = 5$
1,000	$d_{av} = 2.5$ $d_{min} = 2$ $d_{max} = 4$ $p = 0.25$ and $q < 0.1$		$d < 1$ $Ni_{11}Cu_{35}Au_7Sn_{47}$	$d_{av} = 3$ $d_{min} = 2.5$ $d_{max} = 5$
1,500	$d_{av} = 3$ $d_{min} = 2.5$ $d_{max} = 4$ $p = 0.25$ and $q < 0.1$		$d < 1$ $Ni_{23}Cu_{30}Au_5Sn_{42}$	$d_{av} = 3.5$ $d_{min} = 2.5$ $d_{max} = 5$
2,000	$d_{av} = 4.25$ $d_{min} = 3.5$ $d_{max} = 5$ $p = 0.20$ and $q = 0.2$		$d_{av} = 0.75$ $d_{min} = 0.5$ $d_{max} = 1$ $x \sim 0.5$ and $y < 0.1$	$d_{av} = 5$ $d_{min} = 4$ $d_{max} = 5.5$
2,500	$d_{av} = 4.5$ $d_{min} = 3.5$ $d_{max} = 5$ $p = 0.20$ and $q = 0.2$		$d_{av} = 1$ $d_{min} = 1$ $d_{max} = 1.5$ $x \sim 0.5$ and $y < 0.1$	$d_{av} = 5.5$ $d_{min} = 4.5$ $d_{max} = 6$
3,000	$d_{av} = 6.5$ $d_{min} = 6$ $d_{max} = 7$ $p = 0.20$ and $q = 0.2$		$d_{av} = 1.25$ $d_{min} = 1$ $d_{max} = 1.5$ $x \sim 0.5$ and $y < 0.1$	$d_{av} = 7.75$ $d_{min} = 7$ $d_{max} = 8$

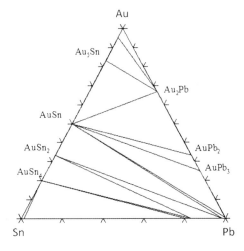

Fig. 6.29 Isothermal section at 150°C from the evaluated Au–Pb–Sn ternary phase diagram

formed. From the Au–Pb–Sn isothermal section at 150°C it can be seen that AuPb-intermetallics can exist in local equilibrium only with almost pure lead, not lead that is saturated with tin. But if the Pb-rich layer becomes almost pure lead (as more and more tin is consumed) during the evolution of the microstructure, as annealing is continued, AuPb-intermetallic compounds can start to form. Before this can take place, one should see the formation of two other AuSn intermetallic compounds, i.e. $AuSn_2$ and AuSn, between the $AuSn_4$ and lead-rich layer as the tin content of the Pb-rich region decreases. This was not detected after annealing up to 3,000 h.

When considering the interface between Ni_3Sn_4 and $AuSn_4$ (Figs. 6.21 and 6.25) the situation is quite similar to above. Based on the ternary Au–Ni–Sn phase diagram determined experimentally at room temperature [78] it is obvious that both phases involved in the above-discussed reaction (Ni_3Sn_4 and $AuSn_4$) can exist in local equilibrium and further that they exhibit extended solid solubilities (Fig. 6.30 and ref [79]).

Based on that what has been stated above, it is not surprising that $(Au,Ni)Sn_4$ redeposits on top of Ni_3Sn_4. This is due to the following reasons: First, $(Au,Ni)Sn_4$ can exist in local equilibrium with Ni_3Sn_4. From Fig. 6.30 it can immediately be seen that in order for local equilibrium to be fulfilled at both interfaces [$Ni_3Sn_4/(Au,Ni)Sn_4$ and $(Au,Ni)Sn_4/Sn$ (Fig. 6.25)], the $(Au,Ni)Sn_4$ must contain the maximum amount of Ni in the Au-sublattice. Under these conditions the Au content of Ni_3Sn_4 is likewise restricted to almost zero. This explains why we could not detect any Au inside Ni_3Sn_4, since only almost pure Ni_3Sn_4, $(Au,Ni)Sn_4$ with a maximum amount of Ni and Sn fulfill the local equilibrium requirement (Fig. 6.30). Second, based on the thermodynamic calculations, we know that Ni has a very strong stabilising effect on $(Cu,Ni)_6Sn_5$ [68].

Thus, it is very likely that Ni also stabilises $(Au,Ni)Sn_4$, since Ni exhibits extensive ternary solubility to $(Au,Ni)Sn_4$. This could act as the driving force for

6.1 Examples of Metal–Metal Interfaces

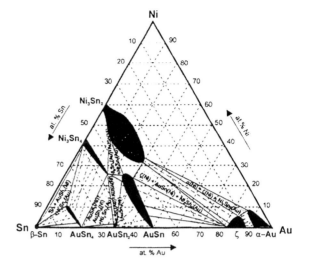

Fig. 6.30 Isothermal section of the Au–Ni–Sn system at room temperature [79] (© 1998 IEEE)

the reaction, since the total Gibbs free energy of the system could be lowered in this way. In order to quantify the second reason, i.e. the effect of Ni on the stability of (Au,Ni)Sn$_4$, a thermodynamic description of the Au–Sn–Ni ternary system should be available. Unfortunately, at the moment this kind of description is unavailable. As Ni$_3$Sn$_4$ forms already during the melting of the solders (Tables 6.1 and 6.3) the formation of the layered structure occurs most likely as follows: during solid-state annealing Au diffuses towards the Ni$_3$Sn$_4$ because or the stabilizing effect of Ni on the (Au,Ni)Sn$_4$. Au reacts with Ni$_3$Sn$_4$ to release Ni that is subsequently incorporated in the growing (Au,Ni)Sn$_4$. Since (Au,Ni)Sn$_4$ must exist in equilibrium at the same time with both Ni$_3$Sn$_4$ and Sn (Fig. 6.30), it will dissolve the maximum amount of Ni into the Au-sublattice.

As has been described, Cu-containing solders do not exhibit the redeposition of (Au,Ni)Sn$_4$ to the interface. In fact some of the available Au is incorporated into the (Cu,Au,Ni)$_6$Sn$_5$ compound (Fig. 6.27). Furthermore, no Ni$_3$Sn$_4$ is formed owing to the reasons discussed above. The formation of (Cu,Au,Ni)$_6$Sn$_5$ also seems to reduce the total intermetallic reaction rate owing to the limited mass-supply of Cu (Table 6.3). It is not immediately clear why the redeposition of AuSn$_4$ does not occur when Cu is present in the solder. One reason may be that Cu$_6$Sn$_5$ cannot exist in equilibrium with the AuSn$_4$. However, from the experimentally determined partial isothermal section of the Au–Cu–Sn systems at 170°C (Fig. 6.31) by Roeder [80] it can be seen that Cu$_6$Sn$_5$ and AuSn$_4$ can in fact exist in local equilibrium and the above explanation cannot be correct.

However, what is different when compared to the local equilibrium established between Ni$_3$Sn$_4$ and AuSn$_4$ is that AuSn$_4$ does not dissolve Cu as it does Ni. This implies that Cu should not stabilize AuSn$_4$ as much as Ni does. Further, one can see that Cu$_6$Sn$_5$ can dissolve extensive amounts of Au. In fact, from the diagram one can see that the local equilibrium between AuSn$_4$ and (Cu,Au)$_6$Sn$_5$ (as well as

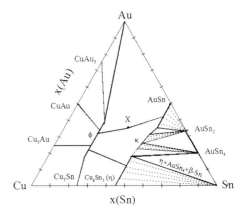

Fig. 6.31 Experimentally determined isothermal section through the Au–Sn–Cu system at 170°C. Redrawn from [80]

with Sn) is possible only if the amount of Au in $(Cu,Au)_6Sn_5$ is close to 20 at.% (Fig. 6.31). Thus, with the small Au layer thickness used in practice the redeposition of $AuSn_4$ should not occur, since it is highly unlikely that such high Au concentrations inside $(Cu,Au)_6Sn_5$ could be realised. Nevertheless, in this investigations there should be enough Au available and the above-described equilibrium should be attainable. The appropriate question would then be *why does $AuSn_4$ not redeposit on the top of $(Cu,Ni,Au)_6Sn_5$?* We propose that this is owing to the following reasons: The limited solubility of Cu in $AuSn_4$ indicates that Cu does not markedly increase the stability of $AuSn_4$. Further, the stabilising effect of Ni on the $(Cu,Ni)_6Sn_5$ ensures that the Ni is strongly bounded in the Cu–Sn–IMC and is not available for $AuSn_4$. The dissolution of Ni (and also probably that of Au) makes the $(Cu,Ni,Au)_6Sn_5$ stable enough that the $AuSn_4$ cannot decompose it and form $(Au,Ni)Sn_4$. Hence, since there is no Ni available at the IMC/solder interface the $AuSn_4$ remains inside the bulk solder. The maximum amount of Au that we found from the $(Cu,Au,Ni)_6Sn_5$ in any of the samples was about 12 at.%. The isothermal section in Fig. 6.30 has been determined at 170°C whereas our annealing were carried out at 150°C. Therefore, the maximum solubility at the annealing temperature should be somewhat smaller than that in Fig. 6.31. Further, since there is also Ni in the $(Cu,Au,Ni)_6Sn_5$ we are actually dealing with quaternary solubility. Thus, the maximum solubility under these conditions may well be the measured 12 at.%. The behaviour of Cu_6Sn_5 is very interesting since it seems to be able to accommodate various species into its structure. This is expected to be related to the fact that Cu_6Sn_5 has an NiAs-based structure that is known to be very flexible and, therefore, able to incorporate both small and large atoms [81].

The (SnAg)/Ni/Au system is thermodynamically quite similar to that of (SnPbAg)/Ni/Au as already discussed. However, the growth rate is significantly lower in the (SnAg)/Ni/Au system (compare the thickness values in Tables 6.1 and 6.3). The $(Au,Ni)Sn_4$ follows parabolic growth kinetics throughout the annealing in the (SnPbAg)/Ni/Au system and up to 500 h of annealing in the (SnAg)/Ni/Au system. When one calculates the parabolic rate constants (k_p) for the two systems for the annealings up to 500 h (annealing period where parabolic kinetics dominates in both systems) one ends up to the result: $k_p^{SnPbAg} \approx 2 \times k_p^{SnAg}$. The appropriate question is then *what causes this difference?* It is known that Au can diffuse rapidly in both Sn and Pb phases via (at least partly) interstitial mechanism [82, 83]. Results of the Au diffusion in Pb and Sn single crystals show that Au diffusion in lead is slightly faster (about five times depending on the Sn crystal axis in question) than that of Au in Sn [82, 83]. This observation, together with the fact that the bulk of the SnPbAg solder contains much more interfacial area (because of the lamellar eutectic structure), which offer fast interfacial diffusion paths for Au, should leads to a higher effective diffusion rate of Au and hence to faster growth rate of (Au,Ni)Sn$_4$ in the SnPbAg than in the SnAg system. Thus, the observed thickness decrease of (Au,Ni)Sn$_4$ in the SnAg solder case after 1,000 h of annealing may be related to the gradually decreasing flux of Au from the solder matrix and subsequent mass-supply problems.

6.1.5 Zn in Lead-Free Soldering

Zn has been suggested to be used in soldering in many ways. Some of the suggested low temperature solders are based on Sn–Zn system [84–90] with additions, such as Bi and Ag. Zn has also been added as an alloying element to SnBi, SnCu and SnAgCu solders [91–95]. There exist also thermodynamic descriptions of the Sn–Cu–Zn system [96–98]. As the amounts of Zn in the above-mentioned applications vary quite much, we shall first consider the Sn–Zn based solder alloys and their reaction with Cu and after that proceed to discuss the role of Zn as a minor alloying element in other Sn-based solders.

The relatively high melting temperature of the Sn–Ag–Cu near-eutectic solder family restricts their use in certain applications and thus solders with lower melting temperature are needed. Among these Sn–Zn solders offer significant benefits on cost as well as on mechanical properties [84]. However, there are major drawbacks with these alloys that include poor corrosion resistance in humid/high temperature environment and poor compatibility with common substrate materials used in electronics (Cu and Ni/Au) [88].

Kivilahti [96] used the assessed thermodynamic description of the quaternary Cu–Sn–Bi–Zn system to investigate the effect of Zn on the wettability of Cu with the eutectic Sn41.5Bi3.5Zn solder. The thermodynamic analysis was performed by combining the available data on the binary Sn–Bi, Sn–Zn, Cu–Bi and Bi–Zn systems [68] and experimental information obtained by the diffusion couple technique. As a result of the assessment the intersection of the CuSnBiZn system was calculated by keeping the Sn-to-Bi ratio at the fixed value of 1.33, i.e. corresponding to the eutectic composition of the SnBi system, at 250°C, so that the plane of intersection joined to the binary CuZn line (Fig. 6.32).

With the Zn-content used, the liquid SnBiZn alloy becomes immediately saturated with Cu (point A) causing the γ-CuZn phase to form. It can be seen from

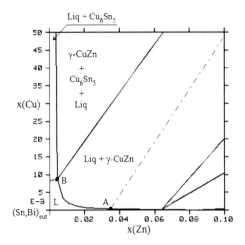

Fig. 6.32 Isothermal section of the SnBiZnCu system at 250 oC with the Sn-to-Bi ratio of 1.33 [96] (© 1995 IEEE)

Fig. 6.32 that the formation of the Cu_6Sn_5—instead of the γ-brass—necessitates the Zn-content to be less than 0.5 at.% (point B). According to the phase diagram the addition of Zn into the Sn–Bi solder prevents the formation of Cu_3Sn completely. Thus, the presence of 41.5 at.% of Bi lowers the critical Zn concentration significantly.

6.1.6 Deformation Induced Interfacial Cracking of Sn-Rich Solder Interconnections

Typically the load-spectrum of a use-environment can be very different for various electronic products. The temperature fluctuations, in any case, are involved in nearly all of them because changes in temperature may be caused either by the external operation environment or by internally generated heat dissipation. The thermomechanical reliability of electronic component boards has therefore been persistently one of the most studied aspects in electronics reliability. Sustained interest in this topic has remained primarily because of (a) the increased power densities and heat dissipation of novel electronic devices and (b) the fact that new materials with different thermomehcanical fatigue characteristics are continuously being introduced into electronics manufacturing processes.

As pointed out in Chap. 3, the reliability of electronic products under thermal cycling conditions is, to a large extent, determined by the ability of solder interconnections to withstand the thermomechanical loads. In spite of being a popular topic of academic research for decades, we are only beginning to understand the complexities related to the failure mechanisms of interconnections under cyclic thermomechanical loading. The reason for this perhaps surprising statement lies in the fact that the much-used SnPb solder alloys were replaced only recently with new lead-free materials and the failure mechanisma of Sn-rich lead-free solder interconnections differs significantly from those of the older lead-containing alloys.

6.1 Examples of Metal–Metal Interfaces

Fig. 6.33 a An optical micrograph obtained with polarised light from a cross-section of the SnAgCu interconnection showing high-angle boundaries between tin grains. **b** An optical micrograph and a superimposed SEM micrograph of the cellular solidification structure

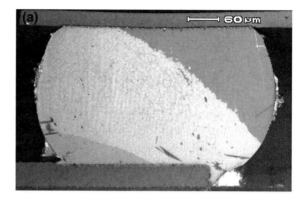

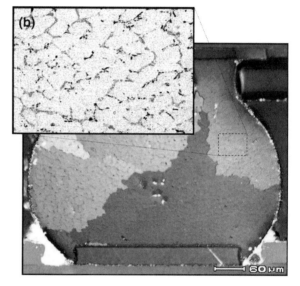

In this case example we discuss the failure mechanisms of lead-free solder interconnections from the perspective of evolution of microstructures, which provide preconditions for the nucleation and propagation of cracks. The microstrucutural approach is particularly important because the failure of solder interconnections is not only a function of load reversals, but many mechanisms related particularly to the evolution of microstructures are dependent on time, temperature, and the amount of deformation. Owing to the extensive research carried out over the last decade we know the microstructures and mechanical properties of many lead-free solder compositions well; the evolution of microstructures during the operation of products, however, has still gained little attention.

As illustrated in Fig. 6.33a, the as-solidified microstructures of Sn-rich solder interconnections are usually composed of relatively few large tin colonies distinguished by high-angle boundaries [99]. The colonies are composed of relatively

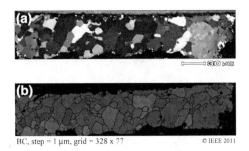

Fig. 6.34 **a** Optical micrograph showing the recrystallised structure on component side interfacial region of the SnAgCu solder interconnection taken with polarised light. **b** EBSD graph of the same location as in (**a**) showing boundaries with large misorientation (larger than 30°C) between the adjacent grains with black lines. Figure is reprinted from [105] with a kind permission of IEEE

uniformly oriented β-Sn cells and the cellular structure of the β-Sn is clearly distinguishable as they are surrounded by thick regions composed of uniformly distributed Cu_6Sn_5 and Ag_3Sn in the β-Sn matrix (see Fig. 6.33b). The microstructural observations of failed solder interconnections, however, indicate that the microstructures of solders are not stable and may change significantly during the operation of products [100–103]. During thermal cycling tests, as-solidified microstructures can transform locally into more or less equiaxed grain structures by recrystallisation. The initiation of recrystallisation is highly localised because only a fraction of the solder interconnection cross-section actually participates in cyclic deformation due to the highly non-uniform distribution of stains and stress inside the solder interconnections. In the process of recrystallisation new strain-free grains are nucleated—after an incubation period—preferentially in the strain concentration regions of solder interconnections at grain or phase boundaries wherefrom they grow in size and consume gradually all the deformed volume of the interconnections [104–106]. The colonies of Sn cells formed during solidification gradually disappear and are replaced by the recrystallised Sn grains.

Figure 6.34 shows a microstructure that in the as-solidified state has looked similar to those in Fig. 6.33 but in the course of thermal cycling has been transformed by recrystallisation. Figure 6.34a shows a micrograph of the recrystallised microstructure on the component side neck region of a thermally cycled interconnection taken with optical microscopy utilising polarised light. Figure 6.34b shows an electron backscatter diffraction orientation map of the same surface. The black lines representing the boundaries where the crystal orientation of the adjacent grains exceeds 30°, correspond well with the grain boundaries visible in the optical micrograph in Fig. 6.34a.

Results published in [100, 107, 108] demonstrate that the cracking of the solder interconnections takes place sometimes before the change of microstructures by

6.1 Examples of Metal–Metal Interfaces

Fig. 6.35 a cross-section of a recrystallised and cracked solder interconnection, (**b**) the same image as in (**a**) but with superimposed crack paths

recrystallisation but the propagation of cracks is sluggish judging by the measured length of cracks without the influence of recrystallisation. However, after initiating the recrystallisation of the solder interconnections, the recrystallised volume is observed to extend gradually over the diameter of the interconnections and intergranular cracking followed the changed volume. Cracks are seldom observed to have propagated further than the recrystallised region. Finally, it is suggested that the networks of the grain boundaries formed by recrystallisation provide favourable paths for cracks to propagate intergranularly (see Fig. 6.35) with less energy consumption compared to the cracking of the solidification colonies.

As discussed in Chap. 3 of this book, a fraction of the energy consumed in the plastic deformation of metals is stored in the form of lattice defects, primarily as dislocations. The increased internal energy of the deformed solder acts as a driving force for the competing restoration processes (i.e. recovery and recrystallisation) and the degree of restoration by recovery depends on the stacking fault energy of the solder alloy. At the time of writing there is a lack of information in the literature about the restoration characteristics of the Sn-based solder alloys but some information on pure Sn can be found. Since most SnAgCu solders contain more than 95 wt% of Sn, recrystallisation studies on pure Sn can be examined indicatively. Creep studies carried out with high-purity tin have indicated that the stacking fault energy of tin would be high [109, 110]. The recovery is very effective in materials with high stacking fault energy due to the efficient annihilation of dislocations. Therefore the restoration of high-Sn solder alloys is expected to take place to a large extent by recovery. Miettinen has shown that that even highly deformed (up to 50% reduction) near-eutectic SnAgCu solders do not recrystallize statically when annealed even at 100°C immediately after the deformation [111]. Also Korhonen et al. [112] failed to observe recrystallisation during room temperature fatigue tests while, on the other hand, they as well as others have demonstrated that near-eutectic SnAgCu interconnections do recrystallise under dynamic loading: thermal cycling as well as power cycling conditions [100–103, 113–118]. This is a further indication that Sn is a high stacking fault energy metal. The decrease of stored energy in high stacking fault energy metals such as Sn takes place very effective by the recovery. Therefore microstructures recrystallise only under restricted loading conditions: dynamic loading condition where strain hardening is more effective than recovery.

A thorough understanding of the restoration process in solders can allow the development of methods for improved lifetime estimation that are based on the evolution of microstructures. Work presented in [104] describes an approach for lifetime prediction based on the competing nature of the restoration processes: Under conditions when strain hardening is more effective than recovery, the cyclic deformation accumulates the stored energy above a critical value after which the recrystallisation can initiate. The total energy of the system consists of the grain boundary energy and the volume stored energy. The stored energy is released through the nucleation and the growth of new grains, which gradually consume the strain-hardened matrix of high dislocation density. Li et al. have developed a multiscale model based on this principle for predicting the microstructural changes of recovery, recrystallisation and grain growth in solder interconnections subjected to dynamic loading conditions [119]. The approach developed in this work is based on the principle that the stored energy of solder is gradually increased during each thermal cycle. When a critical value of the energy is reached, recrystallisation is initiated. Finally, the stored energy is released through the nucleation and growth of new grains, which gradually consume the strain-hardened matrix of high dislocation density. The approach is realised by combining the Monte Carlo (MC) simulations with the finite element calculations. The MC method is employed to model the mesoscale microstructure and the FE method to model the macroscale inhomogeneous deformation. The inhomogeneous volume stored energy distribution in solder interconnections is scaled from the FE model results and mapped onto the lattice of the MC model. The quantitative prediction of the onset of recrystallisation is carried out with the help of the MC simulation. The computational results are compared with the experimentally observed microstructural changes in solder interconnections subjected to thermal cycling tests. This method predicts reasonably well the incubation period and the growth rate of the recrystallisation as well as the expansion of the recrystallised region.

Figure 6.36 shows a comparison of predicted microstructural evolution with experimental evolution. Later on Li et al. have expanded their model to take into consideration the fact that the incoherent primary Cu_6Sn_5 or Ag_3Sn intermetallic particles dispersed in the bulk solder can provide favourable sites for nucleation of recrystallising Sn grains [120]. The onset of recrystallisation is a useful criterion to determine when the material models for the as-solidified microstructures are no longer valid and crack nucleation and propagation should be acknowledged.

6.1 Examples of Metal–Metal Interfaces 177

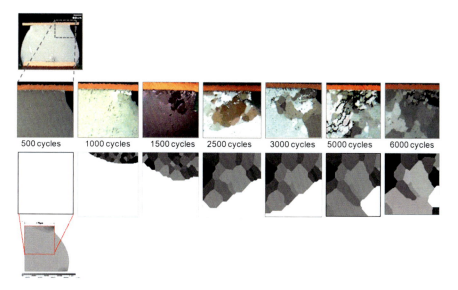

Fig. 6.36 Observed and simulated microstructural changes of solder interconnections with increasing number of thermal cycles

6.1.7 Diffusion and Growth Mechanism of Nb$_3$Sn Superconductor

The bronze technique [97] is one of the most feasible techniques to fabricate multifilamentary wires, in which Nb rods are inserted in Cu(Sn) bronze alloys and the structure is subsequently drawn as a wire. This structure is subsequently annealed at an elevated temperature for a particular time (depending on the requirements) to grow the Nb$_3$Sn superconductor. The overall electrical, mechanical and thermal performance of the structure depends on the morphology that develops during interdiffusion. One of the most important features that is found in the interdiffusion zone is the Kirkendall pores near the (Cu–Sn)/Nb$_3$Sn interface [98]. The position of the pores indicates that Sn is the fastest diffusing species through the Nb$_3$Sn reaction product layer. To confirm this hypothesis, diffusion couple experiments were carried out between Cu-(7–8)at.%Sn and Nb using inert Kirkendall markers [121, 122], as shown in Fig 6.37.

Since the markers were always found at the (Cu–Sn)/Nb$_3$Sn interface, it points out that Sn is virtually the only mobile species through the product layer and the diffusion of Nb must thus be negligible. This is rather surprising when one considers the crystal structure of the Nb$_3$Sn product phase and different kinds of defects (vacancies and antisites) present in the structure [123]. The Nb$_3$Sn phase has an A15 structure, where, as shown in Fig. 6.38a, Sn atoms occupy the body corner and centre positions, whereas, two Nb atoms occupy each face of the cube. The most important feature seen from Fig. 6.38 is that Sn atoms are surrounded by 12 Nb atoms, whereas, as can be seen in Fig. 6.38b Nb atoms are surrounded by

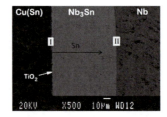

Fig. 6.37 Scanning electron image of the interdiffusion zone, where Nb₃Sn is grown in the Nb/(Cu–Sn) diffusion couple at 775°C annealed for 330 h (Reprinted with permission from [122] © 2010 American Institute of Physics)

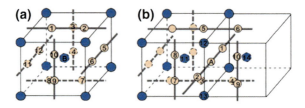

Fig. 6.38 Fig. 2 Nearest neighbours are shown in Nb₃Sn with A15 crystal structure (**a**) Sn atom is surrounded by 12 Nb atoms and (**b**) Nb atom is surrounded by 10 Nb atoms and 4 Sn atoms (Reprinted with permission from [122] © 2010 American Institute of Physics)

10 Nb atoms and 4 Sn atoms. So Nb should be able to diffuse easily via its own sublattice. On the other hand, in a perfect (defect free) crystal, there is no possibility for Sn diffusion, since after the jump Sn would have to sit on the Nb sublattice, which is not energetically favoured. Certain amount of diffusion of Sn is still possible because of the antisite and vacancies present in the structure [123]. Nevertheless, the diffusion rate of Nb should always be higher than Sn in Nb₃Sn compound, if only structural aspects are considered. Thus, the reasons for the observed diffusion behaviour of Sn in the Nb₃Sn product phase needs to be rationalised.

It has been found that with the increase of Sn content just by 1 at.%, as shown in Fig. 6.39 [121, 122], the layer thickness was increased at least by 50%. In addition, there is a large difference in the activation energies for growth (279 kJ/mole in Nb/(Cu-8at.%Sn) and 404 kJ/mole in Nb/(Cu-7at.%Sn) diffusion couple). In order to analyse this, the ternary phase diagram of the Cu–Nb–Sn system was calculated with the help of the assessed data from the Refs. [124–126]. In these calculations, all intermetallic compounds were modelled as binary compounds, since dissolution of ternary elements are found to be negligible based on the available information. In addition, no ternary compounds are known to exist in the Cu–Nb–Sn system. The isothermal section calculated at 700°C is shown in Fig. 6.40.

It is in general consistent with the relatively recent publication on the thermodynamic assessment of the Cu–Nb–Sn system [127], except for two minor details. First, in Ref. [127] the Nb₆Sn₅ phase is stable already below 700°C, different from our calculations and the results of Pan et al. [128]. Second, there is some ternary

6.1 Examples of Metal–Metal Interfaces

Fig. 6.39 a Growth rate of the product phase in Cu(Sn)/Nb diffusion couple for two different atom percentage of Sn, 7 and 8 at.% are shown. **b** Activation energy for the parabolic growth for two different atom percentage of Sn, 7 and 8 at.% are shown (Reprinted with permission from [122] © 2010 American Institute of Physics)

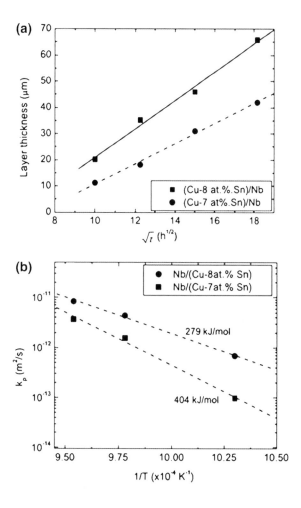

Fig. 6.40 Isothermal section from the Nb–Cu–Sn system at 700°C. The connection line between the two end-compositions of the diffusion couple and the corresponding diffusion path is also shown (Reprinted with permission from [122] © 2010 American Institute of Physics)

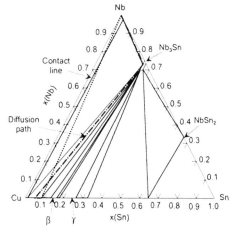

Table 6.4 The difference in chemical potential of elements between two different interfaces as shown in Fig. 6.33 is listed

Temperature (°C)	Sn at.% in Cu(Sn) alloy	$\Delta\mu_{Sn}^{I-II}$ (J/mol)	$\Delta\mu_{Cu}^{I-II}$ (J/mol)	$\Delta\mu_{Nb}^{I-II}$ (J/mol)
700	6	−15,498	508	5,634
	7	−17,391	639	5,765
	8	−19,101	777	6,336
750	6	−13,464	487	4,489
	7	−15,402	621	5,136
	8	−17,150	762	5,719
775	6	−12,226	457	4,076
	7	−14,217	591	4,740
	8	−16,008	733	5,338

Reprinted with permission from [122] © 2010 American Institute of Physics

solubility to both Nb$_3$Sn and Nb$_6$Sn$_5$ in the diagram by Li et al. [127]. However, as the most relevant equilibrium to our analysis is the three-phase triangle Cu–Nb$_3$Sn–Nb, the occurrence of the Nb$_6$Sn$_5$ phase that establishes equilibrium with the liquid phase, has no effect on the equilibria between Nb$_3$Sn and Cu, β or γ phases, even if it is stable already at 700°C or below. To check if there is any effect of the solubility of Cu to Nb$_3$Sn on the relevant phase equilibria, we induced marked ternary solubility to Nb$_3$Sn (up to 10 at.%). The calculations confirmed the fact that this solubility did not have any effect on the three-phase triangle of interest and did not markedly influence the values of chemical potentials of the species. Thus, from the present analysis point of view of these minor discrepancies are acceptable. As discussed earlier, the fundamental fact is that no atom can diffuse intrinsically against its own activity gradient [129]. This condition can be used as a criteria when thermodynamic prerequisites for the diffusion of certain species in a given reaction sequence is rationalised. With the same set of assessed thermodynamic data as used to produce the ternary Cu–Nb–Sn diagram, the values of the driving forces for the diffusion (the difference in chemical potential between two moving interfaces, $\Delta\mu$ in the Cu(Sn)/Nb diffusion couple) of given species over Nb$_3$Sn layer at 700, 750 and 775°C, as a function of Sn content of 6, 7 and 8 at.% in Cu(Sn) alloy were calculated. The results calculated at 775°C are tabulated in Table 6.4.

From the data on the driving force, it is clear that only the diffusion of Sn through the product layer is allowed from the thermodynamic point of view. If Nb or Cu would diffuse intrinsically, they should move against their own activity gradient, which is not possible [132]. Thus, from the thermodynamic point of view there is no possibility of diffusion of Nb in Nb/Cu(Sn) system and the product layer must grow because of diffusion of Sn only. That is why TiO$_2$ particles used as inert markers are found at the Cu(Sn)/Nb$_3$Sn interface.

From the experimental results, we have seen that there is a considerable increase in the growth rate just because of the increase in Sn content from 7 to 8 atomic percent in the Cu–Sn alloy. Figure 6.39 shows the experimental results at 775°C. This difference looks very high considering the minor change in the Sn content. However it does not look surprising, if we consider the change in chemical potential between the two interfaces in the diffusion couple because of the change in Sn content. This increases from −14,217 in Nb/(Cu-7at.%Sn) couple to −16,008 J/mole in the Nb/(Cu-8at.%Sn) at the same temperature. The change is around 12.5%. We can roughly consider that the flux of elements is linearly proportional to the driving force, if we neglect any difference in other parameters because of the change in the composition of Sn in the end member. Thus we can estimate that the amount of Sn flux through the product layer will increase by approximately 12.5%. Moreover, since one mole of Sn produces four moles of the product layer, we can expect that there will be at least a 50% increase in the layer thickness because of the change in Sn content from 7 to 8 at.%. The results presented in Fig. 6.39a support this calculation. So the increase in layer thickness because of such a small change in Sn content in the (Cu–Sn) alloy does not look surprising from the thermodynamic point of view. Since the change in driving force plays an important role in the activation energy, the considerable change in activation energy because of change in Sn content is also rational. Consequently, the high value of activation energy for the growth of Nb_3Sn is also acceptable, since it mainly occurs by diffusion of Sn, which is dependent on the concentration of antisite defects and vacancies present on the Nb sublattice.

6.2 Examples of Metal–Ceramic Interfaces

6.2.1 Cu/diffusion Barrier/Si Systems

The current trend of scaling down the dimensions of integrated circuits in order to achieve better electrical performance places serious demands on the materials used in silicon-based devices. In particular, thin film interconnections are becoming the limiting factor in determining the performance and reliability of integrated circuits. The interconnection delay, usually defined as the RC-delay, where R is the resistance of the interconnection and C is the associated total capacitance, is one of the most important factors determining the device/circuit performance [130]. Reducing the RC-delay to lower than or to equal to the device delay has become both a material and an interconnection design/architecture challenge. Aluminium has been the most widely used material for metallisation in very-large-scale integration (VLSI) and ultralarge-scale integration (ULSI) circuits during the past decades. However, as critical dimensions of devices have approached submicron dimensions, reliability requirements have ruled out the possibility of using pure aluminium. In order to improve reliability, alloys with several additives such as Si, Cu, Ti, Pd, Cr, Mg and Mn, have been tested [131]. The improvement in the electromigration performance thus achieved remains to date limited and is offset by the corresponding increase in interconnection resistivity [131].

Hence, emphasis and burden have been placed on materials e.g., switching from Al/W-based metallisations to Cu- based interconnections. Unfortunately, the interaction between Si and Cu is rapid and detrimental to the electrical

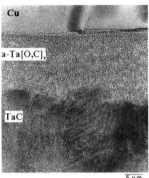

Fig. 6.41 HREM micrograph from the amorphous Ta[O,C]$_x$ phase from the sample annealed at 600°C for 30 min

performance of Si even at temperatures below 473 K [132–140]. The mobility of copper is also relatively fast in SiO$_2$ and many polymers used as dielectric layers [136, 137]. Moreover, copper corrodes easily upon exposure to moisture or oxygen and a technique to passivate the copper surface is essential for multilevel copper interconnection [140]. Therefore, owing to these multiple material problems, it is necessary to implement a barrier layer into the Cu metallisation scheme and to encapsulate copper conductors from all sides. In this respect, diffusion barriers in Cu metallisations differ from the ones used in Al-metallisations, where a diffusion barrier is generally used only at one interface. The need to encapsulate Cu conductor imposes also boundary conditions for diffusion barrier thickness. Among the most promising candidates to be used for this purpose are tantalum-based alloys and binary tantalum nitrides and carbides.

6.2.1.1 Si/TaC/Cu System

In this investigation copper and tantalum carbide films were sputtered onto cleaned and oxide-stripped (100) n-type Si substrates in a dc/rf-magnetron sputtering system. The deposition of TaC was obtained from the TaC-target (hot-pressed, 6.2 wt% of C, main impurity Nb ~ 0.3 wt%) in an argon atmosphere. The pressure before the deposition runs was approximately 10^{-5} Pa. The thickness of the tantalum carbide layers was 70 nm. The copper films with thickness of 400 nm were subsequently sputter deposited without breaking the vacuum. Details of the experimental set-up and analyses can be found from [142].

The as-deposited and films annealed up to 550°C showed clearly discrete layers according to the RBS measurement [141]. After annealing at 600°C the RBS spectrum suggested that the interdiffusion of Ta and Cu had taken place [142]. Investigations with a transmission electron microscope (TEM) revealed that after annealing at 600°C the formation of an amorphous layer at the TaC/Cu interface took place (Fig. 6.41). The composition of the layer was determined to be Ta with marked amounts of oxygen and carbon from the very thin foil (tens of nanometres thick) with the X-ray energy dispersive spectrometry (EDS) in the analytical TEM.

6.2 Examples of Metal–Ceramic Interfaces

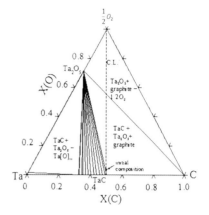

Fig. 6.42 Isothermal section from the evaluated metastable ternary Ta–C–O phase diagram at 600°C under the external oxygen pressure of about 0.2×10^{-4} Pa. The tie-lines in the TaC–Ta$_2$O$_5$ two-phase region are shown in the diagram. The contact line (C.L.) between the TaC film and oxygen indicating the initial unstable equilibrium as well as the approximate composition of the TaC[O]$_{gb}$ are also shown

Equally high amounts of oxygen and carbon were not detected from either side of the amorphous layer. The layer is most probably Ta[O,C]$_x$ (i.e. metastable oxide) containing some carbon released from the partly dissociated TaC layer. The formation of the amorphous layer was most likely caused by the presence of oxygen in the films (mainly at the grain boundaries) and also because of the diffusion of extra oxygen to the films from the annealing environment.

The reasons leading to the formation of the layer can be analysed with the help of Fig. 6.42 that shows the evaluated metastable ternary Ta–C–O phase diagram. The grounds for using the metastable phase diagram instead of the stable one have been previously discussed [141]. As the oxygen content of the TaC films is in the order of 1–2 at.% [142], the initial composition is inside the TaC–Ta$_2$O$_5$–graphite three-phase region. The initial situation, marked as a point on the contact line (C.L.) connecting the end members of the diffusion couple (1/2O$_2$/TaC) in the isothermal section (Fig. 6.42), is highly unstable, since the oxygen inside the TaC film (most probably located mainly at grain boundaries) has been forced there during the sputterdeposition.

The solubility of Ta$_2$O$_5$–TaC is not known, but it is not expected to be substantial at low temperatures. Thus, there must be a large driving force for the formation of Ta-oxide that is in fact more stable than TaC. When the TaC[O]$_{gb}$ films are annealed at elevated temperatures, oxygen starts to dissolve into the TaC matrix, since the solubility is expected to increase as temperature is elevated. This will ultimately result in the formation of the stable Ta$_2$O$_5$ and graphite. However, the kinetics of the system inhibits the formation of stable Ta-oxide (and thereby also graphite, an issue that will be addressed below) at this considerably low temperature and the metastable amorphous layer is formed instead. As the solubilities of impurities to metastable amorphous phases are generally higher than into stable crystalline ones, the amorphous layer can be expected to act as a "drain" for the excess oxygen. At the Si/TaC interface formation of a very thin amorphous layer was observed after annealing at 600°C. Based on the high resolution micrograph shown in Fig. 6.43 the amorphous layer appears to be quite

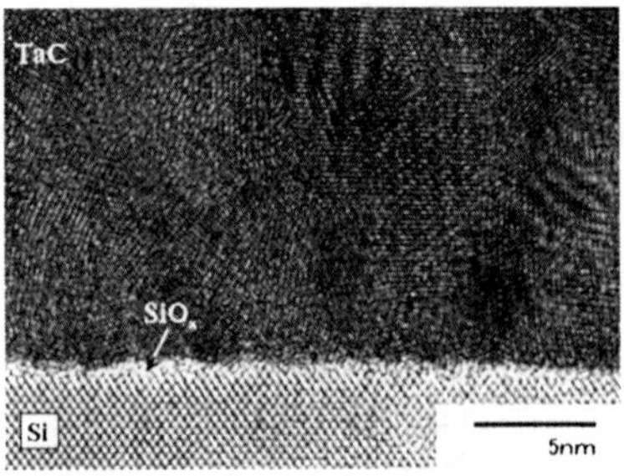

Fig. 6.43 HREM micrograph from the Si/TaC interface from the sample annealed at 600°C for 30 min

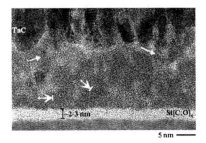

Fig. 6.44 HREM micrograph from the Si/TaC interface from the sample annealed at 750°C for 30 min. *Normal arrows* indicate the crystalline phase and *arrows with a line* at their end indicate the amorphous phase

transparent to electrons and therefore consist of relatively light elements. It is expected that the layer is an amorphous mixture of silicon and oxygen. The reasons for this interpretation are as follows. First, the lower part of the TaC film also contains some oxygen incorporated during the deposition (as shown by EDS elemental mapping). Second, again oxygen has been forced into the TaC layer during processing and the thin film system is far away from equilibrium. As discussed above, the equilibrium solubility of Ta_2O_5 into TaC is not known, but it is not anticipated to be substantial and therefore segregation of oxygen to grain boundaries and interfaces can be expected [142]. This would lead to the accumulation of oxygen at the Si/TaC interface and consequently to the formation of the thin amorphous interphase. In order to verify the composition of the amorphous layer, EDS analyses were carried out. However, as the thickness of the layer is very small (Fig. 6.43) reliable chemical analysis was not possible.

It should be noted that no sputter cleaning was applied to the Si substrate before the TaC film deposition and therefore the "diffuse" interface is not caused by processing. Furthermore, the layer is not expected to be an artefact resulting from sample preparation owing to its visual appearance and moreover, due to the effect of temperature on its thickness as will be discussed below. After annealing at 750°C the amorphous layer at the Si/TaC interface is clearly thicker than after annealing at 600°C (Fig. 6.44). However, the thickness was still so small that it was not possible to reliably analyse the composition of the layer. It is expected that the layer is a $Si[O,C]_x$ type layer i.e. amorphous mixture of silicon, oxygen and carbon and the growth of the layer is due to the stability of stable and metastable Ta-oxides with respect to the TaC as already discussed above in context with

6.2 Examples of Metal–Ceramic Interfaces

Fig. 6.45 Bright field TEM micrograph from the Si/TaC/Cu sample annealed at 750°C for 30 min

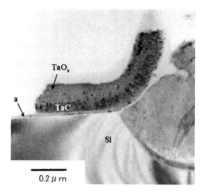

Fig. 6.46 Bright field TEM micrograph from the Si/TaC/Cu sample annealed at 800°C for 30 min

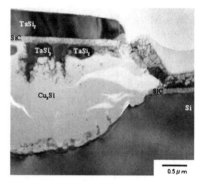

Figs. 6.41 and 6.42. In addition to the growth of the amorphous layer, a mixture of amorphous (arrows with a line in their head) and crystalline phases (normal arrows) can be seen inside the TaC layer at 750°C (Fig. 6.44). The thickness of this phase "mixture" is quite large (~ 10 nm). Figure 6.44 indicates that the phase "mixture" may be growing at the expense of the TaC layer and thus some of the carbon dispersed inside the amorphous layer is anticipated to be expelled from the partly decomposed the TaC layer. Further, the micrograph indicates that there is a boundary between the phase mixture and TaC. This may be quite important from the mechanical point of view as can be seen from Fig. 6.45 that shows a lower magnification micrograph of the whole structure. The ruptured TaC film seen in the micrograph is actually composed of three different layers. From Fig. 6.45 it appears that the bottom layer (the two-phase mixture shown with higher magnification in Fig. 6.44) is still in contact with Si whereas the other two layers have been torn away during sample fabrication. The large Cu_3Si inclusion is also clearly visible on the right hand side.

At 800°C the layered metallisation structure is completely destroyed owing to the extensive interfacial reactions. TaC layer seen at 750°C has been completely consumed at 800°C. Instead, one can see the formation of uniform $TaSi_2$ layer as well as a layer of SiC (Fig. 6.46). The formation of the mixture of amorphous and crystalline layers inside the TaC layer observed at 750°C is expected to be related to the formation of TaSi$_2$ and SiC layers seen in this micrograph. It is probable that the "mixed" layer acted at least partly as a "precursor" for the crystalline TaSi$_2$ and SiC layers. The large Cu$_3$Si "precipitates" seen already at 750°C are also clearly present. There is also considerable amount of oxygen distributed through the layer as shown by the elemental analyses with EDS. This indicates that the Si substrate has been oxidised after the rupture of the overlying films. The thermodynamically expected formation of graphite was not detected even at 800°C. It is anticipated that owing to the very difficult nucleation of graphite the carbon will stay dispersed inside the amorphous phase. There are many reasons for the difficult nucleation of graphite. One probable reason is that the crystal structure is quite complex, and so nuclei do not form easily. Another reason may be that because the coordination number is small, the formation of graphite in a purely solid-state system requires a considerable increase in volume. The strain energy involved in this process makes the nucleation of graphite more difficult, since the mechanical energy associated with the formation of graphite must be extracted from the available free energy driving the nucleation process, as discussed above in Sect. 4.3.5. It is expected that in this system carbon, which is released from the partially decomposed TaC layer and incorporated into the growing amorphous Ta[O,C]$_x$ layer, stabilizes the amorphous structure.

In order to analyse the phase formation at the Si/TaC interface, one should have a firm thermodynamic description available. In principle, to investigate the reactions in the present system, the thermodynamic description of the five-component Si–Ta–C–O–Cu system should be used, since both Cu and O influence the reactions at the Si/TaC interface. Unfortunately, there does not exist adequately reliable data on all the multicomponent systems required. Therefore, the reactions at the Si/TaC interfaces at 800°C are examined with the help of the assessed Si–Ta–C ternary phase diagram together with the calculated activity diagrams (Figs. 6.47 and 6.48), since they already provide a considerable amount of useful information about the interfacial reactions. The main goal is to evaluate the possible reaction sequence at the interface. Hence, the ternary Si–Ta–C phase diagram was evaluated from the assessed binary thermodynamic data [143–145], and is presented in Fig. 6.47. The question of carbon solubility in Ta$_5$Si$_3$, which is expected to be substantial based on the other Me$_5$Si$_3$ silicides (e.g. Ti$_5$Si$_3$ ~ 10 at.%) [146] must be carefully considered in the Si–Ta–C system. However, owing to the lack of reliable ternary data in the literature, ternary solubility has not been incorporated. Possible ternary compounds have similarly been excluded from the diagrams. Assuming that the local equilibrium is achieved at the interfaces of the thin film system, it is possible to use phase diagrams coupled with certain kinetic rules to predict possible or at least to rule out impossible reaction sequences. First, the mass balance requires that the diffusion path, which is a line in the ternary isotherm, representing the locus of the average compositions parallel to the original interface through the diffusion zone [147], crosses the straight line connecting the end members of the diffusion couple at least once. Second, no element can diffuse *intrinsically* against its own activity gradient [129], (i.e. from a low chemical potential area to a high chemical potential area). If such is observed it is induced by the movement of other components.

6.2 Examples of Metal–Ceramic Interfaces

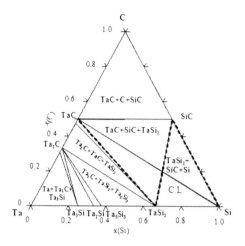

Fig. 6.47 Isothermal section from the evaluated ternary Si–Ta–C phase diagram at 800°C. Contact line between Si and TaC is marked as C.L. The possible diffusion path is also superimposed into the isothermal section

From the phase diagram it is evident that the Si/TaC interface is not in equilibrium and driving force for the formation of additional phases between the substrate and the TaC layer therefore exists (Fig. 6.47). Although there exists a TaC + TaSi$_2$ two-phase region in the phase diagram (Fig. 6.47), SiC must be formed to incorporate the carbon released after the formation of TaSi$_2$ in the reaction between Si and TaC, because of the mass balance requirement. The formation of SiC and TaSi$_2$ was confirmed with TEM investigations and therefore gives support to the assessed phase diagram. As shown in Fig. 6.46 the reacted structure consists of layers of SiC and TaSi$_2$ on top of the silicon substrate. The original TaC is completely consumed during the reaction, since no traces of it can be found at 800°C. The reaction structure seems to be Si, then a layer of SiC, on top of that a layer of TaSi$_2$ and finally on the top layer of TaC (i.e. Si/SiC/TaSi$_2$/TaC), in which the TaC is used completely to yield the final structure Si/SiC/TaSi$_2$. Silicon is expected to be the first species moving at this interface owing to the following reasons. First, Si has been found to be the mobile species during the formation of TaSi$_2$ that occurs around 650°C in the binary Ta–Si system by Si indiffusion whereas the movement of Ta has not been observed under similar conditions [148]. Second, chemical bonding between Ta and C in the TaC compound is expected to be strong, and breaking of these bonds, which is required for the release and subsequent diffusion of Ta, would require large amounts of energy. Owing to the facts stated above, diffusion of tantalum or carbon in this system is not considered to be highly probable. Consequently, Si is anticipated to be the main diffusing species at the Si/TaC interface around 800°C.

Whether or not the above-presented phase formation sequence is thermodynamically possible, can be investigated with the help of Fig. 6.47. The activity diagrams shown in Fig. 6.48 are one form of many different types of stability diagrams. In such a diagram the thermodynamic potential (or activity a_i) of one of the components is plotted as a function of the relative atomic fractions of the other two components. The diagrams have been introduced in Sect. 4.2. It is to be noted

Fig. 6.48 Potential diagrams for Si–Ta–C system at 800°C with the possible diffusion paths superimposed

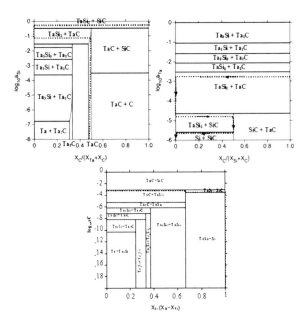

here that inside two-phase areas the diffusion path must follow a horizontal line (tie-line) in order for the local equilibrium to be fulfilled. The activity values can be obtained from the assessed thermodynamic data. As can be seen from the calculated activity diagram, Si can move along its lowering activity gradient in the proposed reaction sequence and therefore diffusion of Si in this particular reaction sequence is allowed on thermodynamic grounds. The examinations of the calculated activity diagrams for carbon and tantalum neither restrict the diffusion of these elements in the suggested reaction sequence (Fig. 6.48). However, as already discussed, carbon and tantalum are strongly bonded to each other in the TaC compound and are not expected to move easily. Therefore, the reaction most likely starts by Si indiffusion into TaC (most probably via grain boundaries). This is followed by the formation of TaSi$_2$, which then leads to the accompanied dissociation of TaC. The released carbon is then available for the formation of SiC in the reaction with Si. This mechanism will finally yield the experimentally observed structure Si/SiC/TaSi$_2$. Thus, carbon and tantalum would not have to move and undergo only local rearrangement. Almost identical behaviour has been observed in a very similar NbC/Si diffusion couple [149]. It must be emphasised than in the present analysis diffusion of oxygen and copper and their influence on the interfacial reactions between Si and TaC have been ignored.

6.2.2 Reactive Brazing of Ceramics

Advanced ceramics are rapidly gaining more importance in modern technology. Complex, multicomponent structures of engineering applications typically require

6.2 Examples of Metal–Ceramic Interfaces

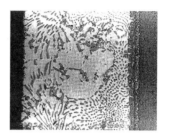

Fig. 6.49 SEM micrograph of Al$_2$O$_3$/steel joint brazed with the Ag27Cu3Ti at 900°C for 5 min

some form of joining. The feasibility of the given structure (typically metal-ceramic) is dictated by the properties of the interface region, which is produced by the chemical reaction between dissimilar materials.

In principal there are various methods to join ceramics to other materials, notably metals. One of the methods providing intimate contact is reactive brazing, either by utilising metallised ceramics or by using active filler alloys. Much research has been focused on the widely used Ag–Cu–Ti system. However, the activation mechanism, i.e. the segregation of Ti at the ceramic interface and the chemical reactions between ceramic and titanium, is still not well understood. The beneficial effect of titanium in the brazing of Al$_2$O$_3$ has been explained, for example, to be due to the formation of TiO$_x$ at the ceramic interface [150]. On the other hand, it has also been reported that on the basis of standard free energy data, the reduction of Al$_2$O$_3$ by Ti would not proceed even at high temperatures [151]. In another study it was shown that depending on the Ag/Cu ratio in the filler alloy, reaction products such as Ti$_3$O$_2$ or (Ti,Cu,Al)$_4$O and [Ti,Al]$_4$Cu$_2$O were found between the Al$_2$O$_3$ and the braze [152]. In addition, the diffusion bonding of Al$_2$O$_3$ to Ti gives rise to an interfacial/reaction layer which can be composed of TiAl, Ti$_3$Al and α-Ti[Al,O] [153].

It will be shown next that by utilising the thermodynamic–kinetic method, a better understanding of detailed mechanisms of active brazing can be achieved. In the experimental setup, sintered alumina (99.7% in purity) was brazed to commercially pure (grade 1) Ti with different Ag–Cu filler alloys. The activity of Ti was varied by changing the Ag to Cu ratio and brazing conditions. The use of bulk Ti makes the starting conditions equal for the different filler alloys. Annealing at 850–950°C for 1–30 min was carried out both in a vacuum furnace as well as in vacuum ampoules. The filler alloy often used in active brazing is an eutectic Ag28Cu(in wt%), which contains a small percentage of Ti. When alumina is joined with Ag27Cu3Ti (in wt%) at 900°C for 5 min the reaction zone consists of an eutectic type structure and the layer next to alumina is 1–2 μm thick (Fig. 6.49). Due to the slow cooling of the sample in Fig. 6.49, one does not obtain information about the evolution of the microstructrure at the bonding temperature. Thus, quenched samples must be utilised for this purpose. The joint between alumina and titanium by using 50Ag50Cu alloy after annealing at 950°C and rapid cooling (quenched in water) is shown in Fig. 6.50. The existence of a miscibility gap in the Ti–Ag–Cu ternary system at the brazing temperature is clearly shown in Fig. 6.50. Liquid L1 (lighter contrast) is Ag- rich and the darker L2 Cu- or Ti-rich.

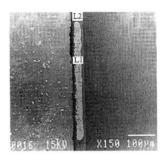

Fig. 6.50 SEM micrograph of Al$_2$O$_3$/Ti joint brazed with 50Ag50Cu at 950°C for 5 min

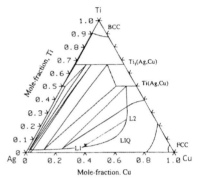

Fig. 6.51 Calculated isothermal section of the Ti–Ag–Cu system at 950°C

It is also to be noted that L1 is enclosed by L2, which forms the actual contact to the ceramic. Thus, the experimental results verify the calculated isothermal section of the Ti–Ag–Cu ternary system shown in Fig. 6.51. When considering the activity of Ti in the two liquids (L1 and L2) it is of course the same. Thus, in principle both liquids should have the same ability to reduce the ceramic. However, the Ti-content in the Cu-rich liquid L2 far exceeds that in the Ag-rich liquid (Fig. 6.51) and therefore only L2 contains enough Ti to form a uniform reaction layer at the ceramic/liquid interface (Fig. 6.50).

How does Ti function in active brazing? To have a more thorough understanding of this issue, Ti–Al–O system should be used (Fig. 6.52). Although titanium produces several stable oxides it distinctively differs from the other reactive metals like aluminium or magnesium by having exceptionally great potential for dissolving large quantities of oxygen. Since titanium dissolves also significant amounts of aluminium, the activities of the component atoms in solid Ti or liquid Ag–Cu–Ti solutions are strongly affected by the composition of the solutions when they are in local equilibrium with the alumina. Previous studies have shown that a saturated solid solution of Ti (i.e. Ti$_2$O) and Ti-aluminides (TiAl and/or Ti$_3$Al) is formed between Ti and Al$_2$O$_3$ in diffusion couples [153, 154].

When pure α- or β-titanium is in contact with the alumina at elevated temperatures it reduces the oxide and dissolves both oxygen and aluminium according to the reaction

6.2 Examples of Metal–Ceramic Interfaces

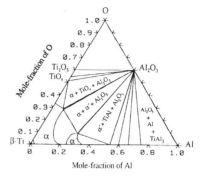

Fig. 6.52 Calculated isothermal section of Ti–Al–O system at 950°C

$$Al_2O_3 \longrightarrow 2[Al]_{Ti} + 3[O]_{Ti} \quad (6.1)$$

Since pure alumina does not dissolve significant amounts of titanium the activities of oxygen (a_O) and aluminium (a_{Al}) in titanium obey the relation

$$K_1 = a_{Al}^2 a_O^3 = \exp(-\Delta°G_1/RT), \quad (6.2)$$

where $\Delta°G_1$ is the standard Gibbs energy of formation of the alumina. During the initial contact between β-Ti and alumina at 950°C, both Al and O dissolve into the titanium, which starts to transform to the α-Ti as soon as the β-Ti is saturated with respect to Al and O (Fig. 6.52). As seen from ternary Ti–Al–O isothermal section at 950°C (Fig. 6.52) α-Ti can dissolve large amounts of Al and O by migrating the α/β boundary away from the original contact interface. Because of the mass balance a ratio of 3 to 2 for [O]/[Al] can be taken for the dissolution of alumina to α-Ti. By utilising the optimised data set used for the calculation of the Ti–Al–O ternary phase diagram, also the metastable solubility line can be evaluated. As discussed in Chap. 4, this gives the maximum amount of Al and O that can be dissolved in α-Ti at any rate. The numerical values ($x_{Al} = 0.143$ and $x_O = 0.215$) are taken from the point where the contact line between Ti and Al_2O_3 cuts the metastable solubility curve. At this point the activities of [Ti], [O], and [Al] are 0.11, 2.48×10^{-18} and 0.09, respectively.

Depending on the activities of the dissolved oxygen and aluminium the reaction between a thin layer of saturated α-Ti[O,Al] and the alumina can produce either oxides (Ti_nO) or intermetallic compounds ($Ti_nAl[O]$). Based on the thermodynamic description of the Ti–Al–O system, it is evident that the formation of oxides is not possible as activity of oxygen should rise at least to the level of 4×10^{-18} for TiO_x or Ti_2O_3 to become stable (Fig. 6.53). As it was seen above, the maximum activity value for oxygen is 2.48×10^{-18}, determined at the metastable solubility. Further, as it is known that the diffusion of oxygen in α-Ti is much faster than that of Al [155] and because the intrinsic diffusion of O is possible only towards its decreasing activity, the local equilibrium cannot change towards the oxygen corner of the diagram. Instead, the local equilibrium moves (along the isoactivity lines) towards the Ti–Al binary system. This makes the α-Ti[Al,O]

Fig. 6.53 Activity of oxygen as a function of a metal ratio $u_{Al}\ [= x_{Al}/(x_{Al} + x_{Ti})]$

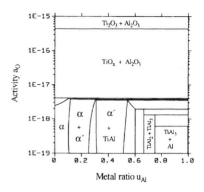

Fig. 6.54 SEM micrograph of the Ti/Al$_2$O$_3$ diffusion couple annealed at 1,000°C 16 h

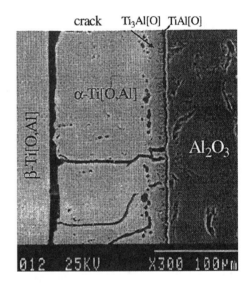

more supersaturated with respect to Ti$_3$Al[O] until the diffusion path has penetrated far enough to the α-α' two-phase field to produce enough driving force for the nucleation of Ti$_3$Al[O] at the α-Ti[O,Al]/Al$_2$O$_3$ interface. Subsequently, Ti$_3$Al[O] becomes also saturated leading to the formation of the TiAl[O] solution phase between Ti$_3$Al[O] and Al$_2$O$_3$ consistent with the experimental observations (Fig. 6.54).

6.2.3 Reactions Between Non-Nitride Forming Metals and Si$_3$N$_4$

Silicon nitride is used in various applications in microsystems. In many cases it is joined with different metals. An interesting manifestation of the role of volatile reaction products during reactive phase formation is found in the case of

6.2 Examples of Metal–Ceramic Interfaces

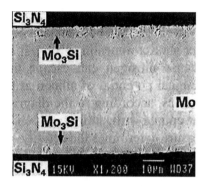

Fig. 6.55 SEM micrograph of the transition zone between dense Si$_3$N$_4$ and Mo after annealing for 32 h at 1,300°C in vacuum [156]

Si$_3$N$_4$/non-nitride forming systems [156]. In this case the type of phases that can form in the diffusion zone (at elevated temperature) between Si$_3$N$_4$ and non-nitride forming metal is found to be different if porous or dense Si-nitride is used. This is because when the interfacial reaction takes place nitrogen gas will form at the interface. Depending on its ability to escape from the interfacial area, wide variety of fugacity values can be reached. To illustrate this, one can examine Fig. 6.51 (from Ref. [156]).

Figure shows phase equilibria in the Mo–Si–N system at 1,300°C under various nitrogen partial pressures (fugacities). When dense Si$_3$N$_4$ is used in the diffusion couple Si$_3$N$_4$/Mo only a layer of Mo$_3$Si is found between Si$_3$N$_4$ and Mo (Fig. 6.55). When one examines the isothermal sections in Fig. 6.56 it is evident that in this case the partial pressure of N$_2$ at the reaction interface must be high. According to Fig. 6.56, it should be between 7 and 88 bar as there is equilibrium between Si$_3$N$_4$ and Mo$_3$Si, but Mo$_2$N has not yet been formed. The high nitrogen partial pressure can be understood by considering the ability of N$_2$ to escape from the reaction zone, when dense Si-nitride is used. In this case it is obvious that the transport of N$_2$ from the reaction zone to the surrounding environment is difficult and a nitrogen pressure build-up will take place. Consequently, the fugacity of N$_2$ in the reaction zone corresponds to that in Fig. 6.56d and results in the experimentally observed phase layer sequence.

On the other hand, when porous Si$_3$N$_4$ is used the N$_2$ released at the reaction interface can easily escape to the surrounding atmosphere and a completely different reaction sequence is found (Fig. 6.57). From the figure it is seen that two Si rich silicides are present in the reaction zone and Mo and Mo$_3$Si, which should have been present during the early stages of annealing, have been completely consumed. It is obvious that in the case of porous Si-nitride (50% porosity), owing to the ability of N$_2$ to escape from the reaction zone no pressure built-up occurs and further that the phase layer sequence is dependent on the external pressure of N$_2$ in the surrounding atmosphere.

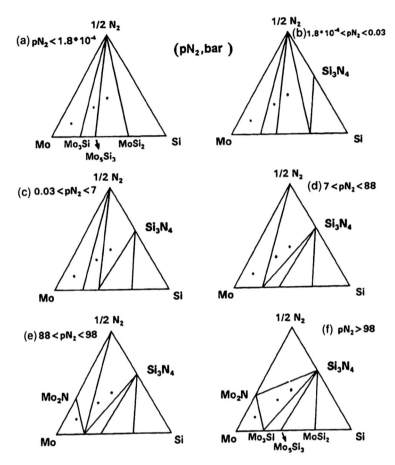

Fig. 6.56 The isothermal cross-sections through the Mo–Si–N system at 1,300°C at various partial pressures of nitrogen [156]

6.3 Example of Metallisation Adhesion on a Polymer Substrate

Polymers have low surface free energy and their reactivity is poor with metals deposited on them. Therefore adhesion in many cases relies on the diffusion of sputter deposited metals into polymer substrates or the mechanical interlocking of chemically deposited metals rather than on significant electrostatic or chemical interactions. One interesting example is the metallisation of unfilled thermosetting epoxy (SU-8) that is increasingly used in MEMS and particularly in bioMEMS due to its biocompatibility. This particular polymer provides high resolution structures in photolithography with ease of processing. This is achieved by not adding any inorganic fillers in the epoxy matrix. This, however causes that the surface of SU-8

6.3 Example of Metallisation Adhesion on a Polymer Substrate

Fig. 6.57 SEM micrograph of the reaction zone between 50% porous Si_3N_4 and Mo annealed for 50 h at 1,300°C [156]

cannot be roughened with wet-chemical or physical plasma etching which in many other cases would provide a reasonable means for mechanical anchorage sites to the adherence of copper [157–162] or nickel [163]. Figure 6.58 presents the effect of RIE treatments on an SU-8 surface.

The smooth and unreactive surface of SU-8 provided inherently ideally poor interfacial adhesion premises for copper adhesion and thus its adhesion improvement methodology is explained next. Wet-chemical oxidation using, for example, $KMnO_4$ does activate the SU-8 surface actually more than physical RIE treatment if the results of improved surface free energy and polarity of the surface are examined [158]. However, wet-chemical treatment does not provide any increase of surface roughness and therefore the use of electroless Cu plating cannot be carried out for SU-8 and physical deposition methods must be employed. Nevertheless, even sputter deposited copper does not adhere to the wet-chemically or reactive ion etched (RIE) surface adequately and the employment of an adhesion promoter layer is needed. Best activation of the SU-8 surface for an inorganic adhesion promoter metallisation of Cr was obtained with the use of a mixture of O_2/CF_4-gas flow in the RIE treatment. It is unclear to us what the exact mechanism between epoxy and Cr is, but others [164] suggest that it is chemical in nature and that the metal would form oxides with the polar groups of the activated epoxy surface. After O_2/CF_4 RIE treatment some fluorine increase together with significant increase in the oxidation of the surface can be seen in XPS analysis, see Fig. 6.59 which suggests that a Cr adhesion promoter could form Lewis acid base complexes at the interface with the treated epoxy surface.

Obviously, if that interface is strong the subsequent copper sputter deposition on the Cr adhesion promoter layer results in good adhesion and it indeed did result in the shift from interfacial delamination to the bulk fracture in the epoxy. In Fig. 6.60 the bulk fracture is within the epoxy is presented. This indicates that the maximum level of adhesion was achieved and the next weakest link in the

196   6 Evolution of Different Types

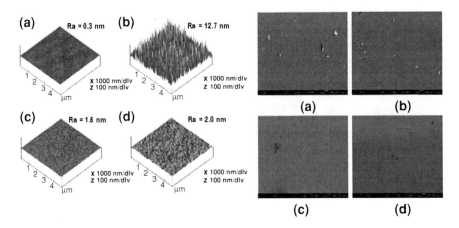

Fig. 6.58 On the *left* AFM and on the *right* SEM micrographs of SU-8 of **a** no treatment; **b** O_2 treatment; **c** CF_4 treatment; **d** combined O_2/CF_4 treatment. Roughness (R_a) is posted on the AFM micrographs. Figure is redrawn from [157]

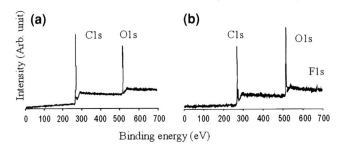

Fig. 6.59 XPS wide scan spectra for SU-8 (**a**) no treatment and (**b**) combined O_2/CF_4 treatment. Figure is redrawn from [157]

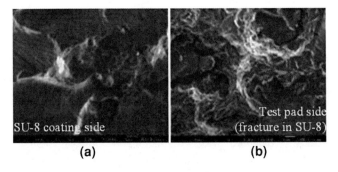

Fig. 6.60 SEM micrograph of O_2/CF_4 treated SU-8/Cr/Cu adhesion system: fracture surface (**a**) SU-8 side and (**b**) backside of the test pad. Representation from [157]

6.3 Example of Metallisation Adhesion on a Polymer Substrate

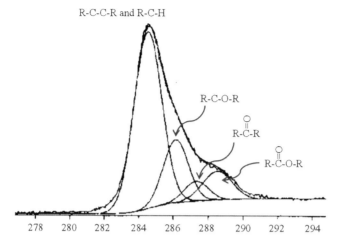

Fig. 6.61 C1 s core level spectrum of the O_2/CF_4 treated SU-8 surface with labels for the characteristic peaks. R in the figure represents $-CH_3$, CH_3CH_2-, C_3H_7-, etc. Redrawn from [157]

Table 6.5 List of some common acids and bases with the categorisation into hard, borderline and soft acids and bases

	Acids	Bases[a]
Hard	H^+, Na^+, K^+	NH_3, RNH_2
	Mg^{2+}, Ca^{2+}	H_2O, OH^-, O^{2-}
	Al^{3+}, Cr^{3+}, Co^{3+}, Fe^{3+}	F^-, Cl^-, NO_3^-, ClO_4^-
	Ti^{4+}, Zr^{4+}	SO_4^{2-}
	Cr^{6+}	CH_3COO^-
Borderline	Fe^{2+}, Co^{2+}, Ni^{2+}, Cu^{2+},	Br^-, NO_2^-, SO_3^{2-}
	Zn^{2+}, Sn^{2+}, Pb^{2+}	
Soft	Cu^+, Ag^+, Au^+	H^-, CN^-, C_2H_4, CO
	Pd^{2+}, Pt^{2+}, Pt^{4+}	RSH, $S_2O_3^{2-}$
	M^0 (metal atoms)	I^-

[a] Represents $-CH_3$, CH_3CH_2-, C_3H_7-, etc

adhesion system appears in the epoxy due to its mechanical properties which cannot be improved by adhesion improvement treatments.

If we take a closer look at the C1s core level spectrum of the O_2/CF_4 treated SU-8 surface, it is found that in addition to the appearance of a fluorine peak there originates a few new carbon oxide groups on the surface. These groups are presented in Fig. 6.61. From Table 6.5 it can be seen that the chromium ion is a strong acid and it is known that such acids like to form chemical bonds with strong bases. Since the adhesion was so significantly improved with the use of a Cr adhesion promoter in particular when the number of fluorine and oxides on the surface increased during RIE treatment, it is reasonable to expect that some of the covalently bonded fluorines and ester groups are dissociated during the Cr deposition enabling the Lewis acid-base complex formation. On the other hand, Cu^+ and Cu^{2+} ions are soft Lewis acids and they do not form complexes with the RIE treated SU-8 surface and even less with the untreated surface. To understand these

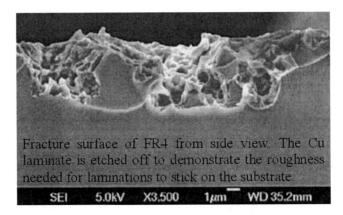

Fig. 6.62 Roughness needed in the substrate to achieve good adhesion without using adhesion promoter. Representation from [158]

Lewis acid–base interactions better and to determine their contribution on the surface free energy of the treated SU-8, the models presented in Chap. 5 were utilised on the next case presented in Sect. 6.4.

6.4 Example of Polymer/Polymerisation Adhesion on Metal Substrate

In many occasions polymers are either polymerised as thermosetting encapsulates, adhesives, prepregs and photoresists or processed as thermoplastic laminates etc. on metals. When a lamination method is used to coat a metal substrate it is usually seemingly rough. The same applies if one laminates on a polymer using metal foils like in the case of printed wiring boards (FR4), see Fig. 6.62. Organic adhesion promoters or more specifically, coupling agents are often used and less roughness is needed as the adhesion mechanism does not have to rely on mechanical interlocking with the use of adhesion promoters but formation of chemical bonding is expected.

The coupling agents are small molecules that are usually two functional. One end of the molecule reacts with a metal surface and the other end points outwards from it reaching towards the polymer. This other end may react with the subsequent polymer layer if the functional group is suitable (carbon–carbon double bond, epoxy, amine etc.) or it may diffuse into the polymer layer if the tail of the coupling molecule is long enough and soluble in the polymer layer forming an interpenetrating network i.e. interphase between the metal substrate and the polymer. Some commonly used coupling agents can be found in literature [165] and in Fig. 6.63 is a presentation of perhaps one of the most common of them. This 3-glysidoxy propyl trimethoxy silane's epoxy end reacts with a thermosetting epoxy prepolymer coating when it is first immobilised on the metal substrate

6.4 Example of Polymer/Polymerisation Adhesion on Metal Substrate   199

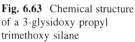

Fig. 6.63 Chemical structure of a 3-glysidoxy propyl trimethoxy silane

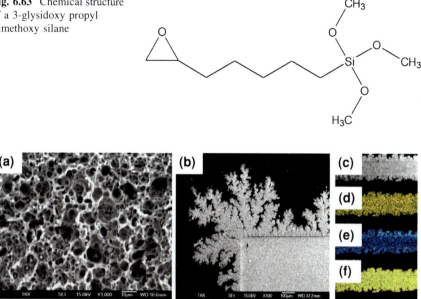

Fig. 6.64 *SEM micrograph of* (**a**) a chemically etched aluminium surface that enables mechanical interlocking with its cavities appearing in micrometre range rather than in nanoscale that was seen in the case 6.X, **b** a electrochemical instability of at a poorly adhered interface, **c** SEM image of a Cu stripe, **d** SEM/EDS Cu map, **e** SEM/EDS Cl map, and **f** SEM/EDS S map

through reaction of three methoxy silane groups. This coupling agent literately links the two materials together. The both ends of the coupling agent may be chosen otherwise to achieve adhesion promotion in other material combinations. Also the length and chemistry of the carbon tail between the two functionalities may be tailored to promote, for example, solubility of it to polymer matrices and thus adhesion through a diffusion mechanism as discussed earlier.

When low viscosity and low filler containing thermosetting polymers are applied on rough surfaces the cavities on the substrate are easily filled and basis for both mechanical interlocking and electrostatic mechanisms is good. Figure 6.64 presents an ideally porous surface for a low viscosity polymer to fill in for mechanical interlocking. This kind of interface may provide good adhesion particularly if it is porous without the use of adhesion promoters but an interface lacking a chemical adhesion mechanism is often very susceptible to the adverse effects of moist environments as is shown in the same figure (b–f) for copper stripes that were exposed to environmental corrosion stress test [166].

When high viscosity adhesives, underfills, encapsulates etc. are applied on rough surfaces there is a high risk of forming air pockets at the interface, see Fig. 6.65. This minimises the contact area with the metal and thus even the mechanical interlocking or electrostatic mechanisms cannot contribute significantly to the adhesion. Also, if coupling agents are not used there may not appear

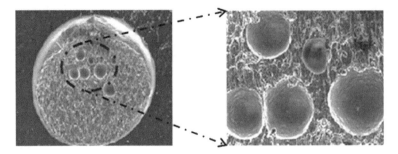

Fig. 6.65 SEM micrographs of the air pockets found at the interface of LCP and silicone adhesive. Redrawn from [157]

adequate amount of reactivity and reasonably viscous epoxy paste may not migrate in the processing temperatures into the metal cavities nor can the metal diffuse into the polymer.

6.5 Example of Polymer Coating Adhesion on a Polymer Substrate

In sensors and bioMEMSs, polymers are often applied on polymer substrates. These polymer couples may be pairs of thermosets, thermoplastics or combinations of the previous. All these cases result in different adhesion features. Figure 6.66 presents schematically these polymer to polymer cases. A special case in which the same polymer material is applied on top of a previous layer of it will be discussed in Sect. 6.6. Here some other polymer to polymer cases are discussed.

When a thermosetting polymer like silicon adhesive is applied on a thermoplastic like liquid crystal polymer (LCP) substrate there are little means to ensure good adhesion. The LCP is inherently known for its non-reactivity and thus chemical stability in various environments. There is no mutual solubility between LCP and silicone so diffusion as an adhesion mechanism is also ruled out. Moreover, the silicone adhesives are often highly filled with inorganic fillers and thus their viscosity is high. This results in kinetic problems. Even if the LCP surface could be treated in a way to provide good wettability, this high viscosity would probably prevent wetting from happening in practise.

Despite these anticipated problems this kind of material interfaces appear in electronics components and they perform adequately. How is this possible? The LCP surface can be chemically activated to some extent using proper oxygen/argon plasma treatment but creation of carbon–carbon double bonds, for example, is difficult for the covalent bonding with silicone adhesive. The number of oxygen-containing groups however can be increased with plasma and this increases the surface free energy of the LCP and thus wettability of it by the silicone adhesive. However, in the studies we did not find any trace of covalent

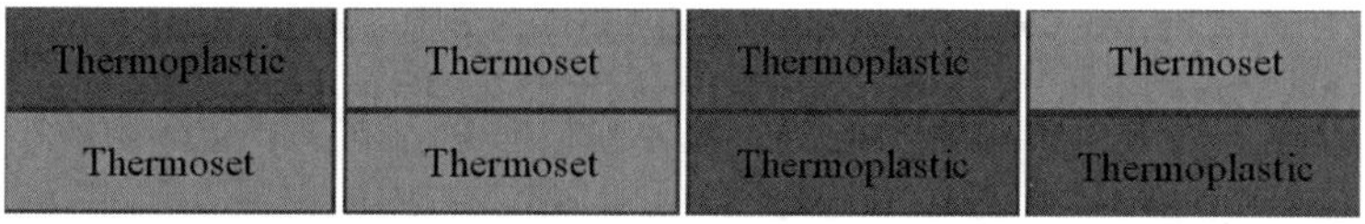

6.5 Example of Polymer Coating Adhesion on a Polymer Substrate 201

| Thermoplastic | Thermoset | Thermoplastic | Thermoset |
| Thermoset | Thermoset | Thermoplastic | Thermoplastic |

Fig. 6.66 Different combinations from the viewpoint of adhesion that can be build up using one thermoplastic polymer and one thermosetting polymer. The one on *top* represents a coating that is applied on the one *below* that represents a substrate

bonding across the interface and it was concluded that the adhesion is based on secondary interactions mainly hydrogen bonding [167, 168]. This provides adequate adhesion if the interface is protected from the penetration of moisture. If water is available it will penetrate to the interface and it is actually thermodynamically favoured because water wants to facilitate those hydrogen bonds between treated LCP and silicone. As more water gets into the interface it starts to act as a lubricant and in adhesion strength tests the interface fails with very low stress levels. We showed with moisture exposures that the adhesion was totally lost in the LCP/silicone case but that it returned to the initial level if the water was removed by baking. This is an example of a polymer–polymer case in which there is a distinct interface between polymer materials and no covalent bonding but only secondary forces contributing to the adhesion.

There are many polymer–polymer cases in which the interface region is actually formed of a reaction product, i.e. interphase of the joint materials or the interdiffusion region of mutually soluble materials at the interface. The mechanical properties of this interphase obviously determine if it will act as a weak boundary layer reducing the adhesion or if it will strengthen the joint. In general, to break the interphase structure, covalent bonds need to be broken whereas in the interdiffusion region it is only needed to overcome the secondary forces between the diffused molecules and the matrix to break the joint. It is to be mentioned that although the last case seems easy to break, the number of secondary interaction sites is so high with macromolecules that they generally contribute to the adhesion an amount comparable to the cohesion fracture of either bulk materials. As mentioned earlier, the coupling agent may penetrate into the polymer in some cases and often thermoplastics when applied on top of another thermoplastic in elevated temperature result in an interdiffusion region.

6.6 Some Examples of Epoxy Polymer Surface Treatments

In the surface treatment of polymers for improved adhesion we basically aim at three goals. First, an attempt is made to increase the surface free energy of polymers by creating polar groups, second, reactive functional groups are desired, and third, roughening or even porosity is looked for. It is important to maximise the surface free energy because that improves interfacial contact and thus electrostatic adherence facilitates mechanical interlocking on rough surfaces. Reactive groups on the surface provide a means to bond covalently to the coating through addition or radical polymerisation and this adhesion mechanism leads to the most reliable joints.

If the use of adhesion promoters and coupling agents is excluded the employment of wet-chemical solutions, physical plasma treatments and mechanical grinding methods in their many variations are available. The last obviously provides a means to create surface roughness and mountain like topography whereas wet-chemical etching may in certain applications lead to even porous surface layers. Creation of porosity requires that the polymer is a composite and that one of the phases can be either etched faster than the matrix polymer or that one of the e.g. inorganic filler phases is removed or dissolved in the treatment. Wet-chemical etching is based on oxidation of the polymer and thus it also creates polar groups on the surface and increases the SFE of the polymer. Plasma treatments may cause slight increase in the roughness but mainly it is used to introduce polar groups on the surface. The used gas in the treatment determines what kind of implantation of functionality can be achieved. Usually oxygen-containing groups are formed but also total coating of the surface with a new material is possible. If the surface for some reason is desired to become non sticky, fluorination may be carried out using double bond bearing fluorinated gases [157–162].

To analyse the changes in the state of the polymer surface the following methods are usually used: (1) instrumental analysis of surface chemistry by X-ray photoelectron spectroscopy (XPS) and surface specific infrared spectroscopy (IR), (2) surface free energy evaluation utilising the theories and models explained in Chap. 5 employing contact angle measurement apparatus, and (3) both optical and electron microscopy methods including surface probing profilometers. These, nowadays widespread methods provide information that is adequate to build a good understanding of the state of the surface and to deduce the prerequisites for adhesion.

The SU-8 epoxy was treated with plasma and wet-chemical methods to improve both copper [157] and epoxy [158] adhesion to it. Simple surface free energy evaluation prior to the coating of SU-8 with either Cu or SU-8 predicted that there are preconditions for good adhesion but it did not provide an adequate picture of the situation since no adhesion between SU-8 and SU-8/Cu was achieved straight away. Therefore XPS analysis was carried out to understand the changes in the surface chemistry. Surface specific IR could be used also but it is to be mentioned that it is probing the surface from deeper (micrometres) than the XPS (nanometres) and indeed IR can show that the treatments cause changes only on the topmost atom layers. The surface appeared exceptionally smooth in microscopy inspections (atomic force microscopy, AFM, surface profilometre analysis, and in scanning electron microscopy, SEM) and there was no reactivity left as seen in XPS analysis. Hence, adhesion promotion treatment was necessary in this case as described earlier. The behaviour of the SU-8 surface under chemical and RIE treatments still puzzled us and therefore, a thorough evaluation of the surface was engaged using many probing liquids and surface free energy calculation approaches as described in [158].

6.6 Some Examples of Epoxy Polymer Surface Treatments

Table 6.6 Surface free energies, $\gamma_{S,GM}$ (mJ/m^2) and polarities, x_p (%) of the 0-, 5-, 10-, 15-, and 20-min treated SU-8s

SU-8	t	$\gamma_{S,GM}$	γ_S^d	γ_S^p	x_p
Untreated		43.6	39.7	3.8	8.7
Chemically	5	44.3	41.5	2.8	6.3
	10	46.1	38.1	8.0	17.4
	15	47.2	38.9	8.3	17.6
	20	47.4	38.5	8.9	18.8
RIE	5	37.9	28.8	9.1	24.0
	10	41.8	24.9	16.9	40.4
	15	41.5	23.7	17.8	42.9
	20	39.5	21.2	18.3	46.3

The surface free energies, dispersion components (γ_S^d), and polar components (γ_S^p) were determined by the geometric mean model (GM) with H$_2$O and diiodomethane as the probing liquids in the contact angle measurements

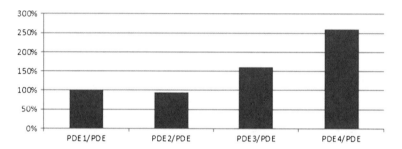

Fig. 6.67 Increase of adhesion in percents (%). Untreated SU-8 (PDE1/PDE), chemically treated (PDE2), RIE treated (PDE3), and not fully crosslinked (PDE4)

This surface free energy investigation revealed that the epoxy surface was chemically very isotropic. In thermoplastics macromolecules may orientate in the direction of melt flow during extrusion but thermosetting prepolymers including epoxies are much smaller and because of their processing steps (described in [158]) there does not appear orientation. We also noticed that RIE treatment increases the polar component of the surface free energy much more than the wet-chemical treatment, see Table 6.6.

This suggests that there were thermodynamically good changes for the complex formation, recall Table 6.5, in the case where the Cr adhesion promoter was sputtered [157] on such RIE treated SU-8 surface. Interestingly, the wet-chemical treatment increased the surface free energy more than the RIE treatment but the polar component remained low. One could quickly deduce that application of SU-8 prepolymer on such crosslinked and treated SU-8 surface would result in improved wettability and thus adhesion. However, the adhesion results presented in Fig. 6.67 show that actually the RIE treatment resulted in slightly better adhesion results. This is because the SU-8 prepolymer is very polar in nature with its many polar unreacted epoxy groups. This kind of prepolymer demands a surface that is indeed polar and the total surface free energy of the surface is not that important. Dramatically this was seen before any treatments when the SU-8 prepolymer did not wet the crosslinked SU-8 surface at all but formation of islands was detected. The thorough surface free energy characterisation provided us with significant information to understand the behaviour of the surface but also lead us to the conclusion that in this particular case, the best way to improve thermoset adhesion to a crosslinked thermoset was to intentionally leave the substrate partially uncrosslinked. To determine the proper degree of crosslinking we studied the crosslinking reaction with dynamic scanning calorimetry (DSC) [169, 170]. When we applied another layer of the polymer on that carefully partially crosslinked film, we achieved seaming adhesion with no interface or interphase as it "fused" into one continuous material.

References

1. T. Uno, K. Tatsumi, Y. Ohno, Void formation and reliability in gold–aluminum bonding. Proc. ASME/JSME Adv. Electr. Pack. EEP **1**(2), 771–777 (1992)
2. T. Uno, K. Tatsumi, Thermal reliability of gold–aluminum bonds encapsulated in bi-phenyl epoxy resin. Microelectron. Reliab. **40**, 145–153 (2000)
3. F. Wulff, C.W. Tok, C.D. Breach, Oxidation of Au4Al in gold ballbonds. Mater. Lett. **61**, 452–456 (2007)
4. M.-H. Lue, C.-T. Huang, S.-T. Huang, K.-C. Hsieh, Bromine-and chlorine-induced degradation of gold–aluminum bonds. J. Electron. Mater. **33**(10), 1111–1117 (2004)
5. S.-A. Gam, H.-J. Kim, J.-S. Cho, Y.-J. Park, J.-T. Moon, K.-W. Paik, Effects of Cu and Pd addition on Au bonding wire/Al pad interfacial reactions and bond reliability. J. Electron. Mater. **35**(11), 2048–2055 (2006)
6. H.-S. Chang, K.-C. Hsieh, Theo Martens and Albert Yang, The effect of Pd and Cu in the intermetallic growth of alloy Au wire. J. Electron. Mater. **32**(11), 1182–1187 (2003)
7. V. Bhatia, F. Levey, C. Kealley, A. Dowd, M. Cortie, The aluminium–copper–gold ternary system. Gold Bull. **42**(3), 201–208 (2009)
8. C. Breach, F. Wulff, An unusual mechanical failure mode in gold ballbonds at 50 µm pitch due to degradation at the Au–Au$_4$Al interface during ageing in air at 175°C. Microeletron. Reliab. **46**, 2112–2121 (2006)
9. C. Xy, C. Breach, T. Sritharan, F. Wulff, S. Mhaisalkar, Oxidation of bulk Au–Al intermetallics. Thin Solid Films **462–463**, 357–362 (2004)
10. J. Poate, Diffusion and reactions in gold films. Gold Bull. **14**(1), 2–11 (1981)
11. I. Belova, G.E. Murch, Diffusion correlation effects and the isotope effect in intermetallic compounds. Phil. Mag. **A80**(9), 2073 (2000)
12. I. Belova, G.E. Murch, A theory of tracer diffusion in non-stoichiometric intermetallic compounds. Phil. Mag **A82**(2), 269 (2002)
13. W.T. Chen, C.E. Ho, C.R. Kao, Effect of Cu concentration on the interfacial reactions between Ni and Sn–Cu solders. J. Mater. Res. **17**, 263–266 (2002)
14. C.E. Ho, R.Y. Tsai, Y.L. Lin, C.R. Kao, Effect of Cu concentration on the reactions between Sn–Ag–Cu solders and Ni. J. Electr. Mater. **31**, 584–590 (2002)
15. C.E. Ho, Y.L. Lin, S.C. Yang, C.R. Kao, D.S. Jiang, Effects of limited Cu supply on soldering reactions between SnAgCu and Ni. J. Electr. Mater. **35**, 1017–1024 (2006)
16. C.E. Ho, Y.L. Lin, S.C. Yang, C.R. Kao, Volume effect on the soldering reaction between SnAgCu solders and Ni, in *Proceedings of 10th International Symposium on Advanced Packaging Materials*, IEEE (2005)

17. E. Ohriner, Intermetallic formation in soldered copper-based alloys at 150–250°C, Weld. J., **191** (1987)
18. A. Paul, in *The Kirkendall Effect in Solid State Diffusion*, Doctoral Dissertation, Technical University of Eindhoven (2004)
19. M. Amagai, A study of nanoparticles in Sn–Ag based lead free solders. Microelectr. Reliab. **48**, 1 (2008)
20. F. Gao, T. Takemoto, H. Nishikawa, Effects of Co and Ni addition on reactive diffusion between Sn-3.5 Ag solder and Cu during soldering and annealing. Mater. Sci. Eng. A **A 420**, 39 (2006)
21. L. Xu, J. Pang, Nano-indentation characterization of Ni–Cu–Sn IMC layer subject to isothermal aging. Thin Solid Films **504**, 362 (2006)
22. V. Vuorinen, T. Laurila, T. Mattila, E. Heikinheimo, J.K. Kivilahti, Solid-state reactions between Cu (Ni) alloys and Sn. J. Electron. Mater. **36**(10), 1355 (2007)
23. P. Oberndorff, *Doctoral Dissertation*, Eindhoven University of Technology, 2001
24. C.-H. Lin, S.-W. Chen, C.-H. Wang, Phase equilibria and solidification properties of Sn–Cu–Ni alloys. J. Electr. Mater. **31**, 907 (2002)
25. C. Schmetterer, H. Flandorfer, Ch. Luef, A. Kodentsov, H. Ipser, Cu–Ni–Sn: a key systemsystem for lead-free soldering. J. Electr. Mater. **38**(1), 10 (2009)
26. K.P. Gupta, An expanded Cu–Ni–Sn system (Copper–Nickel–Tin). J. Phase Equil. **21**(5), 479 (2000)
27. V. Vuorinen, T.M. Korhonen, J.K. Kivilahti, *Formation of the intermetallic compounds between liquid Sn and different Cu–Ni metalization*, Helsinki University of Technology Internal Report, HUT-EPT-4, ISBN 951-22-5629-0 (2001)
28. H. Yu, V. Vuorinen, J.K. Kivilahti, Effect of Ni on the formation of Cu_6Sn_5 and Cu_3Sn intermetallics, in *The Proceedings of the 2006 IEEE/EIA CPMT Electronic Component and Technology Conference (ECTC'06)*, San Diego, California, 29 May–1 June, 2006
29. M. Oh, Doctoral Dissertation, Leigh University, 1994
30. T. Laurila, V. Vuorinen, Combined thermodynamic-kinetic analysis of the interfacial reactions between Ni metallization and various lead-free solders. Materials **2**(4), 1796 (2009)
31. C. Yu, J. Liu, H. Lu, P. Li, J. Chen, First-principles investigation of the structural and electronic properties of Cu6-xNixSn5 (x = 0, 1, 2) intermetallic compounds. Intermetallics **15**, 1471 (2007)
32. F.J.J. van Loo, B. Pieraggi, R. Rapp, Interface migration and the Kirkendall effect in diffusion-driven phase transformations. Acta Metall. Mater **38**(9), 1769 (1990)
33. T.T. Mattila, V. Vuorinen, J.K. Kivilahti, Impact of printed wiring board coatings on the reliability of lead-free chip-scale package interconnections. Mater. Res. **19**, 3214–3223 (2004)
34. F.D.B. Houghton, ITRI project on electroless nickel/immersion gold joint cracking. Circuit World **26**, 10–16 (2000)
35. V.F. Hribar, J.L. Bauer, T.P. O'Donnell, Microstructure of electroless nickel-solder interactions, in *Proceedings of 3rd International SAMPE Electronics Conference*, San Diego, 20–22 June 1989, pp. 1187–1199
36. K. Puttlitz, Preparation, structure, and fracture modes of Pb–Sn and Pb–In terminated flip-chips attached to gold capped microsockets. IEEE Trans. CHMT **13**, 647–655 (1990)
37. E. Bradley, K. Banerji, in *Proceedings of 45th Electronic Components and Technology Conference*, Las Vegas, IEEE, 21–24 May 1995, pp. 1028–1038
38. Z. Mei, P. Callery, D. Fisher, F. Hua, J. Glazer, Brittle interfacial fracture of PBGA packages soldered on electroless nickel/immersion gold, in *Proceedings of Pacific Rim/ASMEInter. Intersociety Electronic and Photonic Packaging Conference*, Advances in Electronic Packaging 1997, ASME, New York, 1997, pp. 1543–1550
39. H.-B. Kang, J.-H. Bae, J.-W. Lee, M.-H. Park, J.-W. Yoon, S.-B. Jung, C.-W. Yang, Characterization of interfacial reaction layers formed between Sn-3.5 Ag solder and electroless Ni-immersion Au-plated Cu substrates. J. Electr. Mater. **37**, 84–89 (2008)

40. V. Vuorinen, T. Laurila, H. Yu, J.K. Kivilahti, Phase formation between lead-free Sn–Ag–Cu solder and Ni (P)/Au finishes. J. Appl. Phys. **99**, 3530–3536 (2006)
41. Z. Mei, M. Kaufmann, A. Eslambolchi, P. Johnson, Brittle interfacial fracture of PBGA packages soldered on electroless nickel/immersion gold. in *Proceedings of 48th Electronic Components and Technology Conference*, Seattle, Washington, IEEE, Piscataway, 25–28 May 1998, pp. 952–961
42. Z. Mei, P. Johnson, M. Kaufmann, A. Eslambolchi, Effect of electroless Ni/immersion Au plating parameters on PBGA solder joint attachment reliability, in *Proceedings of 49th Electronic Components and Technology*, San Diego, IEEE Piscataway, USA, 01–04 Jun 1999, 125–134
43. J.-W. Yoon, S.-W. Kim, S.-B. Jung, S.-B. Jung, Effect of reflow time on interfacial reaction and shear strength of Sn-0.7 Cu solder/Cu and electroless Ni-P BGA joints. J. All. Comp. **385**, 192–198 (2004)
44. S.J. Hang, H.J. Kao, C.Y. Liu, Correlation between interfacial reactions and mechanical strengths of Sn (Cu)/Ni (P) solder bumps. J. Electr. Mater. **33**, 1130–1136 (2004)
45. S.-W. Kim, J.-W. Yoon, Jung S-B interfacial reactions and shear strengths between Sn–Ag-based Pb-free solder balls and Au/EN/Cu metallization. J. Electr. Mater. **33**, 1182–1189 (2004)
46. M. He, Z. Chen, G. Qi, Solid state interfacial reaction of Sn-37Pb and Sn-3.5 Ag solders with Ni-P under bump metallization. Acta Mater. **52**, 2047–2056 (2004)
47. M. He, Z. Chen, G. Qi, C.C. Wong, S.G. Mhaisalkar, Effect of post-reflow cooling rate on intermetallic compound formation between Sn-3.5 Ag solder and Ni-P under bump metallization. Thin Solid Films **462–463**, 363–369 (2004)
48. M. He, A. Kumar, P.T. Yeo, G.J. Qi, Z. Chen, Interfacial reaction between Sn-rich solders and Ni-based metallization. Thin Solid Films **462–463**, 387–394 (2004)
49. M.O. Alam, Y.C. Chan, K.N. Tu, Effect of reaction time and P content on mechanical strength of the interface formed between eutectic Sn–Ag solder and Au/electroless Ni (P)/Cu bond pad. J. Appl. Phys. **94**, 4108–4115 (2003)
50. B.-L. Young, J.-G. Duh, G.-Y. Jang, Compound formation for electroplated Ni and electroless Ni in the under-bump metallurgy with Sn-58Bi solder during aging. J. Electr. Mater. **32**, 1463–1473 (2003)
51. M.O. Alam, Y.C. Chan, K.C. Hung, Reliability study of the electroless Ni–P layer against solder alloy. Microelectron. Reliab. **42**, 1065–1073 (2002)
52. Y.-D. Jeon, K.-W. Paik, K.-S. Bok, W.-S. Choi, C.-L. Cho, Studies of electroless nickel under bump metallurgy—Solder interfacial reactions and their effects on flip chip solder joint reliability. J. Electr. Mater. **31**, 520–528 (2002)
53. P.L. Liu, J.K. Shang, Thermal stability of electroless-nickel/solder interface: part A. Interfacial chemistry and microstructure. Metal. Mater. Trans. A **31A**, 2857–2866 (2000)
54. J.W. Jang, P.G. Kim, K.N. Tu, D.R. Frear, P. Thompson, Solder reaction-assisted crystallization of electroless Ni–P under bump metallization in low cost flip chip technology. J. Appl. Phys. **85**, 8456–8463 (1999)
55. R. Erich, R.J. Coyle, G.M. Wenger, A. Primavera, Solder metallization interdiffusion in microelectronic interconnects, in *Proceedings of 24th IEEE/CPMT International Electronics Manufacturing Technology Symposium*, Austin, IEEE, Piscataway, 18–19 Oct 1999, pp. 16–22
56. S.C. Hung, P.J. Zheng, S.C. Lee, J.J. Lee, The effect of Au plating thickness of BGA substrates on ball shear strength under reliability tests, in *Proceedings of 24th IEEE/CPMT Inter. Electronics Manufacturing Technology Symposium*, Austin, IEEE, Piscataway, 18–19 Oct 1999, pp. 7–15
57. A. Zribi, R.R. Chromik, R. Presthus, J. Clum, K. Teed, L. Zavalij, J. DeVita, J. Tova, E.J. Cotts, Solder metallization interdiffusion in microelectronic interconnects, in *Proceedings of 49th Electronic Components and Technology Conference*, San Diego, IEEE, 1–4 June 1999, pp. 451–457

58. C.W. Hwang, K. Suganuma, M. Kiso, S. Hashimoto, Interface microstructures between Ni–P alloy plating and Sn–Ag–(Cu) lead-free solders. J. Mater. Res. **18**, 2540–2543 (2003)
59. T. Laurila, V. Vuorinen, T. Mattila, J.K. Kivilahti, Analysis of the redeposition of AuSn 4 on Ni/Au contact pads when using SnPbAg, SnAg, and SnAgCu solders. J. Electr. Mater. **34**, 103–111 (2005)
60. T. Laurila, V. Vuorinen, J.K. Kivilahti, Interfacial reactions between lead-free solders and common base materials. Mater. Sci. Eng. R **R49**, 1–60 (2005)
61. T. Laurila, V. Vuorinen, J.K. Kivilahti, Analyses of interfacial reactions at different levels of interconnection. Mater. Sci. Semicon. Process **7**, 307–317 (2004)
62. W.T. Chen, C.E. Ho, C.R. Kao, Effect of Cu concentration on the interfacial reactions between Ni and Sn–Cu solders. J. Mater. Res. **17**, 263–266 (2002)
63. C.E. Ho, R.Y. Tsai, Y.L. Lin, C.R. Kao, Effect of Cu concentration on the reactions between Sn–Ag–Cu solders and N. J. Electr. Mater. **31**, 584–590 (2002)
64. H. Matsuki, H. Ibuka, H. Saka, TEM observation of interfaces in a solder joint in a semiconductor device. Sci. Tech. Adv. Mater **3**, 261–270 (2002)
65. Y.-D. Jeon, K.-W. Paik, K.-S. Bok, W.-S. Choi, C.-L. Cho, Studies on Ni–Sn intermetallic compound and P-rich Ni layer at the electroless nickel UBM-solder interface and their effects on flip chip solder joint reliability, in *Electronic Components and Technology Conference, 2001 Proceedings*, 51st, 29 May–1 June, 2001, pp. 1362–1370
66. R. Ganesan, H. Ipser, Presentation in COST MP 0602 Midterm meeting, Bochum, 15–17 April, Germany, 2009
67. T. Massalski, *Binary Alloy Phase Diagrams*, ASM, (1996)
68. IPMA, *The Thermodynamic Databank for Interconnection and Packaging Materials* (Helsinki University of Technology, Helsinki, 2009)
69. C. Schmetterer, A. Kodentsov, H. Ipser, in *Presentation in COST MP 0602 Midterm meeting*, Bochum, 15–17 April 2009
70. W.J. Johnson, Thermodynamic and kinetic aspects of the crystal to glass transformation in metallic materials. Prog. Mater. Sci. **30**, 81–134 (1986)
71. M.-A. Nicolet, in *Diffusion in Amorphous Materials*, ed. by H. Jain, D. Gupta, TMS, Warrendale, 1994, p. 225
72. A. Minor, J. Morris, Growth of a Au–Ni–Sn intermetallic compound on the solder-substrate interface after agin. Mater. Trans. **31A**, 798–800 (2000)
73. C. Ho, L. Shiau, C. Kao, Inhibiting the formation of (Au $1-x$ Ni x) Sn 4 and reducing the consumption of Ni metallization in solder joints. J. Electr. Mater. **31**, 1264–1269 (2002)
74. L. Shiau, C. Ho, C. Kao, Reactions between Sn–Ag–Cu lead-free solders and the Au/Ni surface finish in advanced electronic packages. Solder. Surf. Mount Tech. **14**, 25–29 (2002)
75. J. Glazer, Microstructure and mechanical properties of Pb-free solder alloys for low-cost electronic assembly: a review. J. Electr. Mater. **23**, 693–700 (1994)
76. K. Zeng, K.N. Tu, Six cases of reliability study of Pb-free solder joints in electronic packaging technology. Mater. Sci. Eng. R **R38**, 55–105 (2002)
77. A. Guy, *Elements of Physical Chemistry* (Addison Wesley, USA, 1960)
78. S. Anhock, H. Oppermann, C. Kallmayer, R. Aschenbrenner, L. Thoman, H. Reichel, Investigation of Au–Sn alloys on different end-metallizations, in *22nd IEEE/CPMT International Electronics Manufacturing Technology Symposium*, Berlin, IEEE, New York (1998) pp. 156–165
79. L. Zavalij, A. Zribi, R. Chromik, S. Pitely, P. Zavalij, E. Cotts, Crystal structure of Au1-xNixSn4 intermetallic alloys. J. Alloys Compounds. **334**, 79–85 (2002)
80. J. Roeder, *Phase Equilibria and Compound Formation in the Au–Cu–Sn System at Low Temperature* (Doctoral Dissertation, Lehigh University, 1988)
81. L. Buene, H. Falkenberg-Arell, J. Gjonnes, J. Gjonnes, J. Tafto, A study of evaporated gold-tin films using transmission electron microscopy: I. Thin Solid Films **67**, 95–102 (1980)
82. B. Dyson, Diffusion of gold and silver in tin single crystals. J. Appl. Phys. **37**, 2375–2377 (1966)

83. C. Kidson, The diffusion of gold in lead single crystals. Phil. Mag. **13**, 247–266 (1966)
84. K. Suganuma, K.-S. Kim, Sn–Zn low temperature solder. J. Mater. Sci. Mater. Electron. **18**, 121 (2007)
85. T.-C. Chang, M.-C. Wang, M.-H. Hon, Effect of aging on the growth of intermetallic compounds at the interface of Sn-9Zn-xAg/Cu substrates. J. Cryst. Growth **250**, 236 (2003)
86. P. Sun, C. Andersson, X. Wei, Z. Cheng, D. Shangguan, J. Liu, Study of interfacial reactions in Sn-3.5 Ag-3.0 Bi and Sn-8.0 Zn-3.0 Bi sandwich structure solder joint with Ni (P)/Cu metallization on Cu substrate. J. Alloy. Compd. **437**, 169 (2007)
87. Y. Liu, J. Wan, Z. Gao, Intermediate decomposition of metastable Cu5Zn8 phase in the soldered Sn–Ag–Zn/Cu interface. J. Alloy. Compd. **465**, 205 (2008)
88. C.-Y. Lee, J.-W. Yoon, Y.-J. Kim, S.-B. Jung, Interfacial reactions and joint reliability of Sn-9Zn solder on Cu or electrolytic Au/Ni/Cu BGA substrate. Microelectr. Eng. **82**, 561 (2005)
89. L. Duan, D. Yu, S. Han, H. Ma, L. Wang, Microstructural evolution of Sn-9Zn-3Bi solder/Cu joint during long-term aging at 170°C. J. Alloy. Compd. **381**, 202 (2004)
90. M. Islam, Y. Chan, M. Rizvi, W. Jillek, Investigations of interfacial reactions of Sn–Zn based and Sn–Ag–Cu lead-free solder alloys as replacement for Sn–Pb solder. J. Alloy. Compd. **400**, 136 (2005)
91. F.-J. Wang, F. Gao, X. Ma, Y.-Y. Qian, Depressing effect of 0.2 wt% Zn addition into Sn-3.0 Ag-0.5 Cu solder alloy on the intermetallic growth with Cu substrate during isothermal aging. J. Electr. Mater. **35**(10), 1818 (2006)
92. X. Chen, M. Li, X. Ren, A. Hu, D. Mao, Effect of small additions of alloying elements on the properties of Sn–Zn eutectic alloy. J. Electr. Mater. **35**(9), 1734 (2006)
93. F. Wang, X. Ma, Y. Qian, Improvement of microstructure and interface structure of eutectic Sn-0.7 Cu solder with small amount of Zn addition. Scr. Mater. **53**, 699 (2005)
94. D.H. DeSousa, L. Patry, S. Kang, D-Y. Shih, The influence of low level doping on the thermal evolution of SAC alloy solder joints with Cu pad structures, in *56th Electronic Components and Technology Conference (ECTC)*, San Diego, 1454, (2006)
95. F.-J. Wang, Z.-S. Yu, K. Qi, Intermetallic compound formation at Sn-3.0 Ag-0.5 Cu-1.0 Zn lead-free solder alloy/Cu interface during as-soldered and as-aged conditions. J. Alloy. Compd. **438**(1–2), 110 (2007)
96. J. Kivilahti, Modelling new materials for microelectronics packaging. IEEE Trans. Compon. Packaging Manuf. Technol. **18**, 326 (1995)
97. A.R. Kaufmann, J.J. Pickett, Multifilament Nb3Sn superconducting wire. Bull. Am. Phys. Soc. **15**, 838 (1970)
98. D.S. Easton, D.M. Kroeger, Kirkendall voids–A detriment to Nb3Sn superconductors. IEEE Trans. Magnetics MAG**15**, 178–181 (1979)
99. T.T. Mattila, T. Laurila, J.K. Kivilahti, Metallurgical factors behind the reliability of high density lead-free interconnections, in *Micro-and Opto-Electronic Materials and Structures: Physics, Mechanics, Design, Reliability, Packaging*, ed. by E. Suhir, C.P. Wong, Y.C. Lee vol. 1, (Springer Publishing Company, New York, 2007), pp. 313–350
100. T.T. Mattila, V. Vuorinen, J.K. Kivilahti, Impact of printed wiring board coatings on the reliability of lead-free chip-scale package interconnections. J. Mater. Res. **19**(11), 3214–3223 (2004)
101. D. Henderson, J.J. Woods, T.A. Gosseling, J. Bartelo, D.E. King, T.M. Korhonen, M.A. Korhonen, L.P. Lehman, E.J. Cotts, S.K. Kang, P. Lauro, D.-Y. Shih, C. Goldsmith, K.J. Puttliz, The microstructure of Sn in near eutectic Sn–Ag–Cu alloy solder joints and its role in thermomechanical fatigue. J. Mater. Res. **19**(6), 1608–1612 (2004)
102. S. Terashima, K. Takahama, M. Nozaki, M. Tanaka, Recrystallization of Sn grains due to thermal strain in Sn-1.2Ag-0.5Cu-0.05 N solder. Mater. Trans. Japan Inst. Met. **45**(4), 1383–1390 (2004)
103. S. Dunford, S. Canumalla, P. Viswanadham, Intermetallic morphology and damage evolution under thermomechanical fatigue of lead (Pb)-free solder interconnections, in *The*

Proceedings of the 54th Electronic Components and Technology Conference, Las Vegas, IEEE/EIA/CPMT 1–4 June 2004, pp.726–736
104. T.T. Mattila, J.K. Kivilahti, The role of recrystallization in the failure mechanism of SnAgCu solder interconnections under thermomechanical loading. IEEE Trans. Compon. Packag. Technol. **33**(3), 629–635 (2010)
105. T.T. Mattila, M. Mueller, M. Paulasto-Kröckel, K.J. Wolter, Failure mechanism of solder interconnections under thermal cycling conditions, in *The Proceedings of the 3rd Electronic Systemintegration and Technology Conference*, Berlin, IEEE/EIA CPMT, 13–16 September 2010
106. H.T. Chen, M. Mueller, T.T. Mattila, J. Li, X.W. Liu, K.-J. Wolter, M. Paulasto-Kröckel, Localized recrystallization and cracking of lead-free solder interconnections under thermal cycling. J. Mater. Res. **26**, 16 (2011)
107. T.T. Mattila, H. Xu, O. Ratia, M. Paulasto-Kröckel, Effects of thermal cycling parameters on lifetimes and failure mechanism of solder interconnections, in *The Proceedings of the 60th Electronic Component and Technology Conference*, Las Vegas, IEEE/EIA CPMT, 1–4 June 2010, pp. 581–590
108. H. Xu, T.T. Mattila, O. Ratia, M. Paulasto-Kröckel, Effects of thermal cycling parameters on lifetimes and failure mechanism of solder interconnections, in *IEEE Transactions on Manufacturing and Packaging Technologies—a Special Issue*, (in print). Invited paper
109. D. Hardwick, C.M. Sellars, W.J. Tegart, The occurrence of recrystallization during high–temperature creep. J. Inst. Met. **90**, 21–22 (1961)
110. D. McLean, M.H. Farmer, The relation during creep between grain–boundary sliding, sub–crystal size, and extension. J. Inst. Met. **85**, 41–50 (1956)
111. S. Miettinen, Recrystallization of lead-free solder joints under mechanical load, Master's Thesis (in Finnish), Espoo, (2005), p. 84
112. T.-M. Korhonen, L. Lehman, M. Korhonen, D. Henderson, Isothermal fatigue behavior of the near-eutectic Sn–Ag–Cu alloy between −25 and 125°C. J. Elec. Mater. **36**(2), 173–178 (2007)
113. P. Limaye, B. Vandevelde, D. Vandepitte, B. Verlinden, Crack growth rate measurement and analysis for WLCSP Sn–Ag–Cu solder joints, in *The proceedings of the SMTA international annual conference*, Chicago, 25–29 Sept 2005, pp. 371–377
114. L. Xu and J. H.L. Pang, Intermetallic growth studies on SAC/ENIG and SAC/CU-OSP lead-free solder joints, Therm Thermomech Phenom Electron Syst (2006), pp.1131–1136
115. U. Telang, T.R. Bieler, A. Zamiri, F. Pourboghrat, Incremental recrystallization/grain growth driven by elastic strain energy release in a thermomechanically fatigued lead-free solder joint. Acta Materialia **55**, 2265–2277 (2007)
116. J. Sundelin, S. Nurmi, T. Lepistö, Recrystallization behaviour of SnAgCu solder joints. Mater. Sci. Eng. A **A 474**, 201–207 (2008)
117. P.T. Vianco, J.A. Rejent, A.C. Kilgo, Time–independent mechanical and physical properties of the ternary 95.5Sn–3.9Ag–0.6Cu solder. J. Elec. Materi. **32**(3), 142–151 (2003)
118. P. Lauro, S.K. Kang, W.K. Choi, D.–.Y. Shih, Effect of mechanical deformation and annealing on the microstructure and hardness of Pb–free solders. J. Elec. Materi. **32**(12), 1432–1440 (2003)
119. J. Li, T.T. Mattila, J.K. Kivilahti, Multiscale simulation of recrystallization and grain growth of Sn in lead-free solder interconnections. J. Elec. Materi. **39**(1), 77–84 (2010)
120. J. Li, H. Xu, T.T. Mattila, J.K. Kivilahti, T. Laurila, M. Paulasto-Kröckel, Simulation of dynamic recrystallization in solder interconnections during thermal cycling. Comput. Mater. Sci. **50**, 690–697 (2010)
121. A.K. Kumar, A. Paul, Interdiffusion and growth of the superconductor Nb_3Sn in Nb/Cu(Sn) diffusion couples. J. Elect. Mater. **38**, 700–705 (2009)
122. T. Laurila, V. Vuorinen, A.K. Kumar, A. Paul, Diffusion and growth mechanism of NbSn superconductor grown by bronze technique. Appl. Phys. Lett. **96**(23), 1910–1912 (2010)
123. R. Besson, S. Guyot, A. Legris, Atomic-scale study of diffusion in A15 Nb {3} Sn. Phys. Rev. B **75**, 0541051–0541057 (2007)

124. M. Hamalainen, K. Jaaskelainen, R. Luoma, A thermodynamic analysis of the binary alloy systems Cu–Cr, Cu–Nb and Cu–V. Calphad **14**(2), 125–137 (1990)
125. C. Toffolon, C. Servant, J.C. Gachon, B. Sundman, Reassessment of the Nb–Sn system. J. Phase Equilibria. **23**(2), 134–139 (2002)
126. J.H. Shim, C.S. Oh, B.J. Lee, D.N. Lee, Thermodynamic assessment of the Cu–Sn system. Z Metallkd. **87**(3), 205–212 (1996)
127. M. Li, Z. Du, G. Guo, C. Li, Thermodynamic optimization of the Cu–Sn and Cu–Nb–Sn systems. J. Alloy. Compd. **477**, 104 (2009)
128. V. Pan, V. Latysheva, Y. Litvinenko, V. Flis, V. Gorskiy, Nb–Cu–Sn system. Phys. Met. Metall. **49**, 170 (1980)
129. F.J.J. van Loo, Multiphase diffusion in binary and ternary solid-state systems. Prog. Solid St. Chem. **20**, 47–99 (1990)
130. C.-Y. Chou, S.-W. Chen, Phase equilibria of the Sn–Zn–Cu ternary system. Acta Mater **5**, 2393 (2006)
131. Y.-C. Huang, S.-W. Chen, C.-Y. Chou, W. Gierlotka, Liquidus projection and thermodynamic modeling of Sn–Zn–Cu ternary system. J. Alloy. Compd. **477**, 283–290 (2008)
132. S.P. Murarka, Advanced materials for future interconnections of the future need and strategy: invited lecture. Microelectron. Eng. **37/38**, 29 (1997)
133. J.T. Yue, Reliability, in *ULSI Technology*, ed. by C.Y. Chang, S.M. Sze (McGraw-Hill, USA, 1996), pp. 656–704
134. J. Torres, Advanced copper interconnections for silicon CMOS technologies. Appl. Surf. Sci. **91**, 112–123 (1995)
135. Broniatowski, Multicarrier trapping by copper microprecipitates in silicon. Phys. Rev. Lett. **62**, 3074 (1989)
136. A.A. Istratov, E.R. Weber, Iron and its complexes in silicon A. Appl. Phys. **66**, 123 (1998)
137. C.-A. Chang, Formation of copper silicides from Cu (100)/Si (100) and Cu (111)/Si (111) structures. J. Appl. Phys. **67**, 566 (1990)
138. J. Li, J.W. Mayer, Refractory metal nitride encapsulation for copper wiring. MRS Bull. **18**, 52 (1993)
139. F. Faupel, Diffusion in non-crystalline metallic and organic media. Phys. Stat. Sol. **134**, 9 (1992)
140. S. Lakshminarayanan, J. Steigerwald, D.T. Price, M. Bourgwois, T.P. Chow, R.J. Gutman, S.P. Murarka, Contact and via structures with copper interconnects fabricated using dual Damascene technology. IEEE Electr. Dev. Lett. **15**, 307 (1994)
141. T. Laurila, K. Zeng, J. Molarius, I. Suni, J.K. Kivilahti, TaC as a diffusion barrier between Si and Cu. J. Appl. Phys. **91**, 5391 (2002)
142. T. Laurila, K. Zeng, J. Molarius, I. Suni, J.K. Kivilahti, Amorphous layer formation at the TaC/Cu interface in the Si/TaC/Cu metallization system. Appl. Phys. Lett. **80**, 938 (2002)
143. J. Lacaze, B. Sundman, An assessment of the Fe–C–Si system. Metall. Trans. A **22**, 2211 (1991)
144. K. Frisk, A.F. Guillermet, Gibbs energy coupling of the phase diagram and thermochemistry in the tantalum-carbon system. J. Alloys Compd. **238**, 167 (1996)
145. L. Chandra Sekaran, SGTE databank, (1987)
146. W. Wakelkamp, *Diffusion and Phase Relations in the Systems Ti–Si–C and Ti–Si–N, Doctoral Thesis, Tech* (University of Eindhoven, Netherlands, 1991)
147. J.S. Kirkaldy, L.C. Brown, Diffusion behaviour in ternary, multiphase systems. Can. Met. Q. **2**, 89 (1963)
148. A. Christou, H.M. Day, Silicide formation and interdiffusion effects in Si–Ta, SiO 2–Ta AND Si–PtSi–Ta thin film structures.Electron. Mat. **5**, 1 (1976)
149. C.R. Kao, J. Woodford, Y.A. Chang, A mechanism for reactive diffusion between Si single crystal and NbC powder compact. J. Mater. Res. **11**, 850 (1996)
150. M.G. Nicholas, D. Mortimer, Ceramic/metal joining for structural applications. Mater. Sci. Tech. **11**, 657 (1985)

References

151. H.H Moorhead, T. Henson, *Ceramic Microstructures'86*, ed. by J. Pask, A. Evans, **21** (1986) pp. 949–958
152. M. Paulasto, J. Kivilahti, Microstructural study of Al2O3/Ti joints brazed with different Ag–Cu filler metals. Ceramic Trans. **35**, 165 (1993)
153. J. Kivilahti, E. Heikinheimo, On the thermodynamics and microstructural evolution of the Ti/Al$_2$O$_3$ diffusion couple, in *Proceedings of 3rd International Conference Joining Ceramics, Glass and Metal*, (ed. by W. Kraft), (1989), pp. 131–138
154. X.L. Li, R. Hillel, F. Teyssandier, S.K. Choi, F.J.J. van Loo, Reactions and phase relations in the Ti–Al–O system. Acta Metall. Mater. **40**, 11 (1992)
155. O. Kubaschewski, O. Kubascehwski-von Goldbeck, P. Rogl, and H.F. Franzen, Titanium: Physico-chemical properties of its compounds and alloys, special issue No. 9, ed. by K. Komarek, International Atomic Energy Agency, Vienna (1983)
156. E. Heikinheimo, A. Kodentsov, J. Beek, J. Klomp, F.J.J. van Loo, Reactions in the systems MoSi3N4 and NiSi3N4. Acta Metall. Mater. **40**, S111–S119 (1992)
157. J. Ge, J. Kivilahti, Effects of surface treatments on the adhesion of Cu and Cr/Cu metallizations to a multifunctional photoresist. J. Appl. Phys. **92**(6), 3007 (2002)
158. M.P.K. Turunen, T. Laurila, J. Kivilahti, Evaluation of the surface free energy of spin-coated photodefinable epoxy. J. Polym. Sci.: Part B. Polym. Phys. **40**, 2137 (2002)
159. J. Ge, R. Tuominen, J. Kivilahti, Adhesion of electrolessly-deposited copper to photosensitive epoxy. J. Adhes. Sci. Technol. **15**(10), 1133–1143 (2001)
160. J. Ge, M.P.K. Turunen, J. Kivilahti, Surface modification of a liquid-crystalline polymer for copper metallization. J Polym Sci: Part B. Polym Phys **41**, 623 (2003)
161. J. Ge, M.P.K. Turunen, J. Kivilahti, Surface modification and characterization of photodefinable epoxy/copper systems. Thin Solid Films **440**(1–2), 198 (2003)
162. J. Ge, M.P.K. Turunen, M. Kusevic, J. Kivilahti, Effects of surface treatment on the adhesion of copper to a hybrid polymer material. J Mater Res **18**, 2697–2707 (2003)
163. T.F. Waris, M.P.K. Turunen, T. Laurila, J. Kivilahti, Evaluation of electrolessly deposited NiP integral resistors on flexible polyimide substrate. Microelectronics Reliability **45**(3–4), 665–673 (2005)
164. L.J. Martin, C.P. Wong, Chemical and mechanical adhesion mechanisms of sputter-deposited metal on epoxy dielectric for high density interconnect printed circuit boards. IEEE Trans. Comp. Packag. Tech. **24**(3), 416 (2001)
165. K.L. Mittal, *Silanes and Other Coupling Agents*, vol. Vol. 5 (Brill Academic Publishers, Netherlands, 2009), p. 450
166. M.P.K. Turunen, P. Marjamäki, M. Paajanen, J. Lahtinen, J. Kivilahti, Pull-off test in the assessment of adhesion at printed wiring board metallisation/epoxy interface. Microelectron. Reliab. **44**, 993 (2004)
167. Tölö, *Adhesion between a LCP package and a silicone adhesive, Master's Thesis* (Aalto University, Finnish, 2011), p. 50
168. S. Niiranen, *Deterioration of Adhesion Between Plasma Treated LCP and Silicone Under Accelerated Environmental Stress Tests, Master's Thesis* (Aalto University, Finnish, 2011), p. 63
169. M.P.K. Turunen, T. Laurila, J. Kivilahti, Reactive blending approach to modify spin-coated epoxy film: part II. Crosslinking kinetics. J. Appl. Polym. Sci. **101**(6), 3689 (2006)
170. M.P.K. Turunen, T. Laurila, J. Kivilahti, Reactive blending approach to modify spin-coated epoxy film: part I. Synthesis and characterization of star-shaped poly (ϵ-caprolactone). J. Appl. Polym. Sci. **101**(6), 3677 (2006)

Index

A
Acid-base complex, 198
  acid-base model, 107
  acid component γ^+, 26, 110
Activation energy, 29, 48, 58, 89, 90–92, 180, 182
Activity coefficient, 50, 54, 55, 58
Activity diagram, 77, 82, 83, 187, 189
Activity gradient, 47, 77, 80, 81, 83, 181, 187, 189
Activity, 47, 49, 50, 52, 54, 55, 58, 67, 75–83, 121, 128, 140, 141, 157, 181, 187–193, 196, 200, 201, 203
Adhesion, 3, 5, 10, 102, 103, 105–109, 111, 113–134, 136, 141, 196–205, 212
Adhesion promoter, 121, 123, 197, 198–200, 203, 204
Adhesion test, 124–126, 128, 129, 133
  Adsorption mechanism, 119
Air pocket, 200, 201
  atomic force microscopy (AFM)
  base component γ^-, 110
  blister test, 128
Alloy, 2, 10, 15, 16, 19, 21, 28, 29, 33, 42, 48, 52, 55, 58–61, 63, 68, 69, 77, 78, 89, 95, 98–100, 121, 131, 136, 141, 142, 145–148, 157, 159, 160, 172, 173, 176, 177, 181–183, 190, 205–212
  binary, 45, 49, 53, 57, 61
  ternary, 22, 45, 53, 61, 68, 69, 71
Amorphous films, 89
Antisite, 179, 182
Au–Al–Cu system, 7o, 137, 138, 140
Au–Ni–Sn system, 74, 169

Au–Pb–Sn system, 64
Au–Sn–Cu system, 74, 145

B
Bga (ball grid array), 153
Binary, 20, 45, 49, 50, 53, 57, 58, 61–63, 65, 67, 69, 71, 72, 77, 80, 81, 83, 98, 99, 111, 157, 158, 172, 179, 183, 187, 188, 192, 208, 211
Binding energy, 53
Blister test, 129
Bolzmann-matano, 76
Bulk fracture, 126, 197, 8, 33, 37, 53, 71, 79, 81, 83–86, 88, 89, 91–93, 95, 97, 102, 103, 105, 107, 116, 120, 124–126, 129, 150, 157, 158, 160, 163, 171, 172, 177, 190, 197, 202, 205
  capillary action, 112
  Cassie-baxter equation (chemical heterogeneity of the surface)
  cavitation test, 112
  centrifugal test, 128
Chemical adsorption, 102, 103

C
Calphad method, 68, 69, 93, 95, 98
Chemical potential, 46, 47, 49, 50, 51, 60, 61, 72, 75–79, 81, 88, 149, 150, 181, 187
Chemical reaction, 45, 51, 71, 104, 132, 190
Combined thermodynamic-kinetic method, 4, 71

C (*cont.*)
Common tangent, 51, 60, 61, 72, 79, 95
Component, 1, 6–9, 12–14, 16–22, 24–27, 30–35, 37, 39, 40, 42, 45–50, 53, 55, 57, 60, 61, 63, 67, 69, 71, 72, 75, 77, 78, 80–82, 88, 94, 95, 99, 100, 102, 105, 107, 108, 110, 111, 116, 117, 128–130, 132, 145, 152, 158, 159, 173, 175, 187–189, 191, 202, 204, 206–210
Concentration, 31, 40, 55, 63, 76, 78, 80, 83, 85, 86, 91, 95–97, 99, 100, 105, 128, 142, 146, 148, 150, 159, 160, 171, 173, 175, 182, 205, 208
Concentration gradient, 76, 78, 95–97, 100
Configuration entropy, 50
Connector (electromechanical), 22
Contact angle, 107
  contact angle measurement, 202
Coupling agent, 115, 198
  critical contact angle, 107
Critical Cu content, 142, 144
Critical gradient, 97
Critical size nucleus, 92
Cu–Sn–Bi–Zn system, 171

D
Delamination, 10, 102, 103, 124–126, 129, 133, 197
  diffusion mechanism, 119, 120, 200
  dipole induced forces (Debye), 118
  dipole-dipole forces (Keesom), 118
  dispersion component γ^P, 109
  dispersion forces (London force), 119
Die attach, 2, 10, 11, 15
Die bond, 10
Die mount, 10
Diffusion, 3–5, 8, 12, 45, 47, 48, 61, 71–73, 76–89, 91–95, 98–100, 102, 115, 118–125, 127, 137, 140, 141, 144, 146–151, 156, 159, 160, 162, 164, 172, 177–184, 184, 187–194, 196, 200–202, 205–212
  bulk, 86, 88, 95
  grain boundary, 30, 53, 84, 87
  L
  short circuit, 83, 84, 87
  volume, 32, 84, 91
Diffusion barrier, 81, 83, 89, 99, 100, 146, 159, 182, 183, 211
Diffusion coefficient, 76, 78, 79, 84, 85, 99, 146, 150, 151
  integrated, 7, 19, 181

interdiffusion, 76, 78, 94, 114
intrinsic, 58, 77, 155
tracer, 78, 79
Diffusion couple, 71, 78, 80–83, 86, 99, 100, 144, 147–149, 172, 178–181, 184, 187, 189, 191, 193, 210, 212
Diffusion path, 80–83, 86, 87, 140, 148, 149, 172, 180, 188, 189, 193
Driving force, 29, 45, 51, 52, 61, 69, 72–77, 79, 81, 87, 89–95, 97, 116, 141, 149–151, 169, 181, 182, 184, 188, 193
Dynamic scanning calorimetry (DSC), 205
  electrostatic mechanism, 118, 199
  enthalpy of adhesion
  epoxy
  equation-of-state model
  excess free energy

E
Effect of Ni on Cu–Sn reactions
Electromigration, 8, 77, 182
Embedded components, 18, 19
Encapsulation, 15, 24, 115, 211
Enthalpy, 46, 49, 50, 55, 103
Entropy, 46, 47, 49, 50
Environment, 1, 6, 10, 11, 15–17, 19, 24, 25, 30, 32, 33, 39, 41, 42, 46, 47, 84, 102, 103, 106, 123–125, 127, 129, 133, 140, 141, 172, 173, 184, 195, 200, 201, 212
Equilibrium
  complete
  global
  local
  metastable
  partial
  pressure
Electroless Ni/immersion Au (ENIG), 152
Eutectic bonding, 16
Eutectic reaction, 61, 63
Eutectoid reaction

F
Fick's first law, 76
Fick's second law, 76
Filler, 24, 116, 118, 125, 190, 196, 200, 201, 203, 212
Fisher's model
Flexibilizer
Flip-chip, 13, 14, 22
Free volume, 124, 125

Index

geometric mean model
harmonic mean model, 107, 110
hydrogen bonding, 109, 118, 201
Fusion bonding, 16, 190

G

Gibbs energy diagram, 50, 71, 79, 88
Gibbs free energy, 46, 47, 49–51, 68, 71, 72, 75, 96–98, 105, 151, 170
Glass bonding, 10
Glass frit bonding, 16
Grain boundary, 29, 53, 71, 84–89, 99, 146, 147

H

Henry's law, 55, 57
Hydrogen bonding, 119, 202

I

Ideal solution, 50, 55
Impurities, 11, 83, 84, 86, 87, 93, 136, 155, 184
Infrared spectroscopy (IR), 203
Integrated, 6, 7, 18, 79, 182
Interaction parameter, 54–56, 69, 145
Interchange energy, 54
Interdiffusion, 45, 71, 76, 78–80, 93, 94, 98, 99, 115, 120, 150, 151, 178, 183, 202, 207, 210, 211
Interfacial adhesion, 102, 103, 105–107, 109, 111, 113, 115, 117, 119, 121, 123–129, 131, 133, 136, 196
Interfacial interactions, 120
Interfacial energy, 97
Interpenetrating network, 199
Intrinsic, 12, 57, 58, 77, 78, 81, 114, 115, 150, 156, 181, 187, 192
Isolated, 46
Isopleth, 65, 67
Isothermal section, 63–68, 70, 82, 83, 138–140, 142, 148, 149, 163, 169–171, 173, 179, 180, 184, 188, 191, 192, 194

K

Kelvin equation, 105, 106
Kinetics of wetting, 103, 113–116
Lap joint, 125, 129
Kirkendall effect, 206
Known good die, 15

L

Laplace equation, 105
LCP, 123, 127, 129, 133, 201, 202, 212
LED, 1–3, 8–11, 13, 15–17, 34, 40, 49, 50, 53, 55, 59, 63, 71, 76–78, 80, 81, 83, 84, 86, 88, 89, 93–95, 102, 103, 105, 107, 109, 110, 115, 116, 117, 123, 125, 127, 129, 131, 133, 141, 144, 146, 154, 156–160, 162, 167, 169, 175–179, 182–187, 189, 190, 193, 195, 196, 200, 201, 203, 204
Lever rule, 59, 63
Lewis acid-base components γ^{AB}, 110
Lewis acid-base interaction i.e. Lifshitz van der waals interactions, 197
Lewis acid, 109, 111, 119, 197, 198
Lewis base, 109, 119
Lifshitz/van der waals interactions γ^{LW}
Liquid crystal polymer (LCP), 201
Mechanical interlocking mechanism, 119, 120
Local, 2, 4, 10–12, 14, 22, 33, 40, 45, 47, 48, 52, 69, 71, 81, 83, 95, 96, 124, 138–140, 142, 143, 145, 149, 157, 158, 163, 167, 169, 170, 175, 187, 189, 191, 192, 210

M

Mass balance, 81, 187, 188
Metastable, 3, 45, 48, 49, 52, 68, 69, 71, 73, 74, 83, 89, 92, 94, 98, 143, 148, 149, 157, 158, 184, 185, 192, 209
Microscopy, 29, 46, 126, 175, 203, 208
optical, 4, 10, 16, 21, 29, 46, 115, 174, 175, 203
peel test, 125, 126, 133
physical adsorption, 103, 104
plasma etching, 132, 196
polar component γ^P, 110, 203
polarity χ^P, 195
polyethylene, 134
polyimide, 6, 118, 132, 212
Microstructure, 3, 4, 20, 25, 28, 29, 33, 39, 41, 42, 46, 48, 49, 84, 85, 87, 89, 91, 93, 95, 97, 154, 169, 174, 175, 177, 206–210, 212
Miscibility gap, 55, 149, 158
Mixing, 49, 50, 53–55, 95
Molding, 140
Monotectic reaction, 63
Monotectoid reaction, 63
Mo–Si-n system, 193
Multimaterial system, 1–5

N

Nanocrystalline nisnp, 156, 158, 159
Nb–Cu–Sn system, 179, 210
Nb$_3$sn superconductor, 177, 178, 209
Nearest neighbors, 53
Nernst–Einstein relation, 76, 78
Non-uniform system, 96
Nucleation, 39, 40, 41, 48, 74, 80, 89–95, 97, 100, 174, 177, 187, 193
   heterogeneous, 6, 49, 52, 89, 114, 118, 131
   homogeneous, 47, 49, 58, 78, 80, 88, 95, 118, 119, 177

O

Open, 28, 43, 46, 49, 118, 152
Ostwald's rule, 48, 71

P

Parallel tangent construction, 88
Partial, 2, 46–48, 55, 61, 88, 124, 170, 187, 194, 195, 205
Penetration depth, 79
Perfect solution, 55
Peritectic reaction, 61, 63
Phase, 3, 4, 21, 22, 28–30, 42, 45–74, 76–83, 85–100, 105, 106, 109, 110, 113, 116–121, 126, 127, 130, 133, 137, 138, 140–143, 146, 148–150, 155–159, 162–166, 168, 169, 172, 173, 175, 179–181, 183–189, 192–195, 199, 202, 203, 205–209, 211, 212
Phase diagram, 45, 47, 49, 50, 52, 53, 61–63, 65–71, 76, 78, 80, 81, 86, 94, 98, 138, 140, 142, 143, 148, 157, 163, 169, 173, 179, 184, 187, 188, 192, 208, 211
Phase rule, 52, 53, 67, 80, 109, 130
Poisson ratio, 93
Polymer, 1, 5–7, 10, 22, 24, 26, 27, 29, 102, 103, 105–125, 127, 129–133, 136, 137, 140, 183, 196, 198 –205, 212
   pull out
   pull-off, 125, 126, 128, 133, 212
   reactive ion etching (RIE), 118
   reversibility, 104
   Scanning acoustic microscopy (SAM)
   Scanning electron microscopy (SEM), 46
   scratch test
Pressure, 2, 10, 11, 16, 46, 47, 49, 51, 52, 55, 57, 58, 60, 66–69, 104–106, 109, 127, 128, 130, 183, 184, 194, 195
Printed circuit board, 17, 133, 212
Printed wiring board, 18, 24, 30, 33, 34, 124, 133, 142, 152, 199, 206, 209, 212
Property, 47, 71, 72, 113, 128
   extensive
   intensive, 46
   Molar, 46, 49, 54, 61, 71, 72, 75, 88, 98, 100, 106, 211
   partial
Purple plague, 11, 13

R

Raoult's law, 55, 57, 58
Reaction kinetics, 3, 146
Redeposition of AuSn$_4$, 154, 160, 164
Regular solution, 53–55, 69, 70
Rozeboom diagram, 63, 64

S

Sauer-freise analysis, 76
Secondary forces/secondary bonding, 102
   shear test, 125, 129
   silicone, 116, 127, 129, 132, 133, 201, 202, 212
   specifity, 103
   spreading, 113–116, 130–132
   stud pull, 127
   Su-8, 5, 17, 118, 196–199, 203–205
Segregation, 71, 84, 87, 88, 99, 153, 185, 190
Si–Ta–C system, 186
Si–Ta–Cu system, 67, 68
Silicone, 115, 122, 126, 128, 184, 187, 192
Snagcu solder, 28, 38, 42, 145, 152, 154, 157, 172, 175, 176, 205, 208, 210
Sn–Cu–Ni system
Solid-state diffusion, 71, 156
Solubility, 49, 50, 63, 72–74, 78, 87, 118, 120, 122, 123, 130, 138–140, 143, 148, 149, 154, 157, 169, 171, 179, 181, 184, 185, 187, 192, 200, 201
   equilibrium, 29, 45, 46–49, 51–53, 57, 58, 60, 61, 63, 64, 67–73, 73, 77, 81–83, 87–91, 94–96, 99, 105, 106, 107, 113, 138–143, 145, 148, 149, 157, 158, 163, 167, 169–171, 179, 181, 184, 185, 189, 191, 192, 194
Solute, 27, 48, 55, 87
Solution, 1, 2, 6, 8, 9, 15, 21, 42, 49, 50, 53–58, 63, 69, 70, 72–76, 93–95, 98, 115–118, 121, 123, 130, 138, 142, 143, 146, 148, 149, 154–157, 163, 171, 179, 191–193, 196, 203

Index

Solvent, 55, 87, 115
Space model, 63
Standard state, 50, 51, 58, 78
Steep concentration gradients, 95
Strain energy, 91–93, 187, 210
Stress and strain analysis, 3
Surface free energy, 91, 106–111, 113, 116, 126, 130–132, 196, 199, 202–205, 212
Surface modification, 116, 118, 121, 131, 132, 212
Surface roughness, 16, 17, 113, 132, 196, 203
Surface stress, 107
Surface tension, 91, 98, 105–111, 115, 116, 130, 132
System, 1–9, 11, 13, 15, 17–19, 21, 24, 45–49, 51–53, 55, 58, 61, 63–72, 74–78, 80–83, 86–89, 92–96, 98–100, 102, 103, 105, 107, 109, 111, 113–117, 119, 121, 123, 125–133, 136–139, 141, 145–147, 149, 154, 157–161, 163, 164, 166, 167, 170–173, 177, 179–185, 187–194, 197, 205, 206, 208, 210–212

T

Tape automated bonding, 12
Tape test, 125
Ternary, 21, 45, 51, 53, 61, 63–71, 80, 81, 83, 94, 99, 138–140, 148, 149, 155–158, 160, 163, 169, 170, 172, 179, 181, 184, 187, 188, 190–192, 205, 210, 211
Thermodynamic factor, 78
Thermodynamic-kinetic method, 4, 5, 46, 48, 50, 52, 54, 56, 58, 60, 62, 64, 66, 68, 70–72, 74, 76, 78, 80, 82, 84, 86, 88, 90, 92, 94, 96–98, 100
Thermoplastic, 122, 123, 199, 201, 202, 204
Thermosetting, 15, 115, 121, 123, 124, 127, 196, 199–201, 204
Tracer, 78, 79, 205
Trial and error method, 2–5

U

Under bump metallurgy (UBM), 141
Underfill, 15, 115, 132, 200

V

Vacancy, 79, 150
Viscosity, 115, 116, 128, 132, 200, 201
Volume, 6, 9, 14, 18, 27, 28, 31, 47, 50, 53, 69, 72, 84, 85, 88, 90–97, 106, 124, 125, 141, 175–177, 187, 205
Wafer bonding, 16, 17
Wafer level packaging, 15
Wet-chemical etching, 203
Wettability, 106, 114–116, 131, 133, 172, 201, 202, 204
Wetting, 10, 103, 106, 108–110, 112–116, 123, 131, 146, 152, 201
Wire bonding, 10, 11, 15, 24, 136, 137
Work of adhesion, 102, 126–128, 133